If You're Trying to
Get Better Grades
&
Higher Test Scores
in Science,
You've Gotta Have This Book !

Grades 6 & Up

by Imogene Forte
& Marjorie Frank

Incentive Publications, Inc.
Nashville, Tennessee

Illustrated by Kathleen Bullock
Cover by Geoffrey Brittingham
Edited by Jean K. Signor & Charlotte Bosarge

ISBN 0-86530-589-7

3   4   5   6   7   8   9   10   06   05   04

PRINTED IN THE UNITED STATES OF AMERICA
www.incentivepublications.com

Contents

GET SHARP . . . on Earth Science

 101

GET SHARP . . . on Physical Science   181

✔ Get Ready ⟶

Get ready to get sharper in science. Get ready to be a better student and get the grades you are capable of getting. Get ready to feel better about yourself as a student. Lots of students would like to do better in school. (Lots of their parents and teachers would like them to also!) Lots of students CAN do better. But it doesn't happen overnight. And it doesn't happen without some thinking and trying. So, are you ready to put some energy into getting more out of your learning efforts? Good! The first part of getting ready is wanting to do better—motivating yourself to get moving on this project of showing how smart you really are. The **Get Ready** part of this book will help you do just that: get inspired and motivated. It also gives you some wonderful and downright practical ways to organize yourself, your space, your time, and your science homework. Even more than that, it gives tips you can use right away to make big improvements in your study habits.

✔ Get Set ⟶

Once you've taken a good, hard look at your goals, organization, and study habits, you can move on to other skills and habits that will get you set up for more successful learning. The **Get Set** part of this book starts out with a crash mini-review of some tools you'll need for doing science lessons and investigations. Then it gives an overview of the thinking skills that will help you get the most out of your brain. Top this off with a great review of skills you need for good studying. It's all right here at your fingertips— how to listen and read carefully, take notes, study for tests, and take tests. Take this section seriously, and you'll start making improvements immediately in your science performance.

✔ Get Sharp ⟶

Now you're ready to mix those good study habits and skills with the content that you need to learn. The **Get Sharp** sections of this book contain all kinds of facts and explanations, processes and definitions, lists and how-to information. These sections cover the basic areas of science that you study—the nature of science, science history, the big ideas and processes that underlie all branches of science, Earth science, space science, life science, and physical science. The pages of the **Get Sharp** sections are loaded with the information you need to understand science homework and get it done right. You will find this part of the book to be a great reference tool PLUS a *how-to-manual* for many science topics and assignments. Keep it handy whenever you do any science assignment. And don't overlook the **Get Sharp on Science Terms** section of the book. It is a smashingly complete glossary of most of the science terms you'll ever need. Use it to keep those definitions clear!

How to Use This Book

Students

This can be the ultimate homework helper for your science instruction and assignments. Use the *Get Ready* and the *Get Set* sections to improve your attitude and organization for study and to sharpen your study skills. Then, have the book nearby at all times when you have science work to do at home, and use the *Get Sharp* sections to . . .

. . . reinforce a topic you've already learned.

. . . get different and fresh examples of something you've studied.

. . . check up on a fact, definition, process, or detail of science.

. . . get a quick answer to a science question.

. . . get clear on something you thought you knew but now aren't sure about.

. . . guide you in science processes (such as how to do a science investigation).

. . . check yourself to see if you've got a fact or process right.

. . . review a topic in preparation for a test.

Teachers

This book can serve multiple purposes in the classroom. Use it as . . .

. . . a reference manual for students to consult during learning activities or assignments.

. . . a reference manual for yourself to consult on particular facts, processes, and concepts.

. . . an instructional handbook for particular science topics.

. . . a complete glossary of science terminology. (See *Get Sharp on Science Terms*.)

. . . a remedial tool for anyone needing review of a particular science process or concept.

. . . a source of advice for parents and students regarding homework habits.

. . . an assessment guide to help you gauge student mastery of science processes or skills.

. . . a source of good resources for making bridges between home and school. *(Use the letter on page 17 and any other pages as take-home pieces for parents.)*

Parents

The *Get Ready* and *Get Set* sections of this book will help you to help your child improve study habits and sharpen study skills. These can serve as positive motivators for the student while taking the burden off you. Then, you can use the *Get Sharp* sections as a source of knowledge and a process guide for yourself. It's a handbook you can consult to. . .

. . . refresh your memory about a science concept, process, or fact.

. . . get a clear definition of a science term. (See *Get Sharp on Science Terms*.)

. . . end confusion about concepts, processes, facts, formulas, classifications, and many other science questions.

. . . provide useful homework help to your child.

. . . reinforce the good learning your child is doing in science class.

. . . gain confidence that your child is doing the homework right.

GET READY →

Get Motivated

Dear Student,

Nobody can make you a better student. Nobody can even make you WANT to be a better student. But you CAN be. It's a rare kid who doesn't have some ability to learn more, do better with assignments and tests, feel more confident as a student, or get better grades. YOU CAN DO THIS! You are the one (the only one) that can get yourself motivated.

Probably, the first question is this: "WHY would you want be a better student?" If you don't have an answer to this, your chances of improving are not so hot. If you do have answers, but they're like Carla's (page 15), your chances of improving still might be pretty slim. Carla figured this out, and decided that these are NOT what really motivate her. Now, we don't mean to tell you that it's a bad idea to get a good report card, or get on the honor roll, or please your parents. We're not trying to say that getting into college is a poor goal or that there's anything wrong with getting ready for high school either.

But—if you are trying to motivate yourself to be a better student, the reasons need to be about YOU. The goals need to be YOUR goals for your life right now. In fact, if you are having a hard time getting motivated, maybe it is just BECAUSE you're used to hearing a lot of "shoulds" about what other people want you to be. Or maybe it's because the goals are so far off in some hazy distant future that it's impossible to stay focused on them.

So it's back to the question, "Why try to be a better student?" Consider these as possible reasons why:

- to make use of your good mind (and NOT short-change yourself by cheating yourself out of something you could learn to do or understand)

- to get involved—to change learning into something YOU DO instead of something that someone else is trying to do TO you

- to take charge and get where YOU WANT TO GO (It's YOUR life, after all)

- to learn all you can for YOURSELF—because the more you know, the more you think, and the more you understand—the more possibilities you have for what you can do or be in your life RIGHT NOW and in the future

Follow the "Get Motivated Tips" on the next page as you think about this question. Then write down a few reasons of your own to inspire you toward putting your brain to work, showing how smart you are, and getting even smarter.

Sincerely,

Imogene and Marjorie

Get Motivated

1. Think about why you'd want to do better as a student.

2. Think about what you'd gain now from doing better.

3. Get clear enough on your motivations to write them down.

4. Set some short-term goals *(something you can improve in a few weeks)*.

5. Think about what gets in the way of doing your best as a student.

6. Figure out a way to change or eliminate something that interferes. *(Use the form on page 16 to record your thoughts and goals.)*

Why should I be a better student ?

To please my parents
To please my teachers
To impress other kids
To impress my parents' friends
So people will like me better
To keep from embarrassing my parents
To do as well as my older brother
To do better than my sister
So teachers treat me better
To do as well as my mom or dad did in school
To get the money my parents offer for good grades
To get well-prepared for high school
To make a lot of money when I finish school
To get a good report card
To get into college

None of these really motivate me much at all.

Get Ready Tip # 1

Set realistic goals. Choose something you actually believe you can do. And, you'll have better chance of success if you set a short time frame for a goal.

I can do this!

Why do I want to be a better student? What difference would it make for me, now and in the future?
(Write a few reasons.)

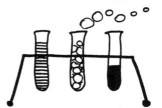

1._____

2._____

3._____

What changes could I make in the near future?
(Write two short-term goals—things that, realistically, you could improve in the next month.)

1._____

2._____

What gets in the way of good grades or good studying for me?
(Name the things, conditions, or distractions that **most often** keep you from doing your best as a student.)

1._____

2._____

3._____

4._____

What distraction am I willing to eliminate ?
(Choose one of the interferences above that you'd be willing to try changing or getting rid of for the next month.)

1._____

Dear Parent:

What parent doesn't want **his/her** child to be a good student? Probably not many! But how can you help yours get motivated to do the work it takes? You can't do it for her (or him), but here are some ideas to help students as they find it within themselves to get set to be good students:

Read the letter to students (page 14). Help your son or daughter think about where she/he wants to go, what reasons make sense to her or him for getting better grades, and what benefits he/she would gain from better performance as a student.

Help your child make use of the advice on study habits. (See pages 18-28.) Reinforce the ideas, particularly those of keeping up with assignments, going to class, and turning in work on time.

Provide your child with a quiet, comfortable, well-lit place that is available consistently for study. Also provide a place to keep materials, post reminders, and display schedules.

Set family routines and schedules that allow for good blocks of study time, adequate rest, relaxing breaks, and healthy eating. Include some time to get things ready for the next school day and some ways for students to be reminded about upcoming assignments or due dates.

Demonstrate that you value learning in your household. Read. Learn new things. Show excitement about learning something new yourself. Share this with your kids.

Keep distractions to a minimum. You may not be able to control the motivations and goals of your child, but you can control the telephone, computer, Internet, and TV. These things actually have on-off switches. Use them. Set rules and schedules.

Help your child gather resources for studying, projects, papers, and reports. Try to be available to take her or him to the library, and offer help tracking down a variety of sources. Try to provide standard resources in the home (dictionaries, thesaurus, computer encyclopedia, etc.).

DO help your student with homework. This means helping straighten out confusion about a topic (when you can), getting an assignment clear, discussing a concept or skill, and perhaps working through a few examples along with the student to make sure he/she is doing it right. This kind of involvement gives a chance to extend or clarify the teaching done in the classroom. Remember that the end goal is for the student to learn. Don't be so insistent on the student "doing it himself" that you miss a good teaching or learning opportunity.

Be alert for problems, and act early. Keep contact with teachers and don't be afraid to call on them if you see any signs of slipping, confusion, or disinterest on the part of your child. It is easier to reclaim lost ground if you catch it early.

Try to keep the focus on the student's taking charge of meeting his/her own goals rather than on making you happy. This can help get you out of a nagging role and get some of the power in the hands of the student. Both of these will make for a more trusting, less hostile relationship with your child on the subject of school work. Such a relationship will go a long way toward supporting your child's self motivation to be a better student.

Sincerely,

Imogene and Marjorie

Get Organized

Abby knows a lot about the solar system. She has done some good research. But she's not really able to show what she's learned, because she is so disorganized. Don't repeat Abby's mistakes.

Get Your Space Organized

Find a good place to study. Choose a place that . . .

. . . is always available to you.

. . . is comfortable.

. . . is quiet and as private as possible.

. . . has good lighting.

. . . is relatively uncluttered.

. . . is relatively free of distractions.

. . . has a flat surface large enough to spread out materials.

. . . has a place to keep supplies handy.
   (See page 19 for suggested supplies.)

. . . has some wall space or bulletin board space for posting schedules and reminders.

Get Ready Tip # 2
Set this up before school starts each year. Make it cozy and friendly—a safe refuge for getting work done. Put a little time into making it your own, so it's a place you like —not a place to avoid.

Get Your Stuff Organized

Gather things that you will need for studying or for projects, papers, and other assignments. Keep them organized in one place, so you won't have to waste time running around looking. Here are some suggestions:

Handy Supplies
- a good light
- a clock or timer
- bulletin board or wall
  (for schedule & reminders)
- pencils, pens, erasable pens
- erasers
- colored pencils or crayons
- markers, highlighters
- notebook paper, typing paper
- scratch paper
- drawing paper
- index cards, sticky notes
- poster board
- folders, report folders
- protractor, compass
- ruler, measuring tape
- calculator
- meter stick, yard stick
- tape, scissors
- glue, glue sticks
- paper clips, push pins
- stapler, staples
- standard references:
  - science textbook
  - encyclopedia (set or CD)
  - homework hotline numbers
  - homework help websites
  - good science websites
- optional science equipment:
  - magnets
  - microscope
  - telescope
  - chemistry set

Get Ready Tip # 3
Have a place to put things you bring home from school. This might be a shelf, a box, or even a laundry basket. Put your school things in there every time you come in the door—so important stuff doesn't get lost in the house or moved or used by other family members.

Get set with a place to keep supplies.
(a bookshelf, a file box, a paper tray, a drawer, a plastic dish pan, a plastic bucket, a carton, or plastic crate)

- Keep everything in this place at all times.
- Return things to it after you use them.

Also have:
- an assignment notebook (See page 22.)
- a notebook for every subject
- a book bag or pack to carry things back and forth
- a schedule for your week (or longer)

Better Grades & Higher Test Scores / SCIENCE
Copyright ©2003 by Incentive Publications, Inc., Nashville, TN.

Get Your Time Organized

It might be easy to organize your study and space and supplies, but it is probably not quite as easy to organize your time. This takes some work. Here's a plan you can follow right away to help you get your time organized.

Think about how you use your time now.

1. For one week, stop at the end of each day, think back over the day, and write down what you did in each hour-long period of time for the whole day.

2. Then look at the record you've kept to see how you used your time. Ask yourself these questions:

Did I have any clear schedule?
Did I have any goals for when
I would get certain things done?
Did I even think ahead about how
I would use my time?
How did I decide what to do first?
Did I have a plan or did I just get things
done in haphazard order?
Did I get everything done or did I run out of time?
How much time did I waste?

8:00 am
I left my science assignment on the breakfast table.

10:30 am
I crammed for my science test between English and gym classes.

3:30 pm
I hung out with friends until dinner.

5:00 pm
I talked on the phone for 2 hours.

7:00-8:00 pm
I watched TV. I thought it would warm up my brain.

9:00 pm
I started my science homework.

9:10 pm
I fell asleep about half way through the chapter on energy.

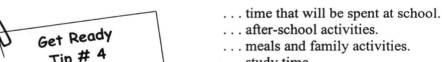

Get Ready Tip # 4

When you plan your week's schedule, don't make it too tight or too rigid. **Leave room for unexpected events.**

3. Next, start fresh for the upcoming week. Make a plan. Include. . .

. . . time that will be spent at school.
. . . after-school activities.
. . . meals and family activities.
. . . study time.
. . . fun, sports, or recreational activities.
. . . social activities or special events.
. . . time for rest and sleep.

4. Make sure you have an assignment notebook. When you plan your weekly schedule, transfer assignments from that notebook into your study time. *(Did you leave enough time to do all these assignments?)*

5. Make a Daily *To-Do* List *(For each day, write the things that must be done by the end of that day.)*

Mon. TO DO List
Study for Math test.
Finish English short stories.
Review for Grammar quiz.
Finish History Ch 11.
Schedule weekend study time.
Wash tennis clothes.
Return library books.
Work on Spanish essay.
Check Internet for info
on Cold War.

Wed. TO DO List
Finish History timeline.
Read Health, Ch 8.
Math problems, pg 221.
Work on autobiography.
Get biography at library.
Start Eng. Ch 8.

Get Ready Tip # 5
At all times—keep a copy of class outlines, schedules, or long-range class assignments at home.

M 5th	T 6th	W 7th	TH 8th	F 9th
8 am–3 pm School Due: Science project Math: pg 215 Health: pg 78	Math test, Ch 7 Due: English: 3 short stories History: Read Ch 12	Grammar quiz Due: Spanish essay Math: pg 219 (even # problems)	Due: History timeline Health: Ch 9 Math: pg 221	3 pages max! Due: English: autobiography English: Ch 8 myths
4–7 pm tennis practice relax dinner	tennis match 4 pm relax dinner	tennis practice relax dinner	tennis practice relax dinner wash tennis clothes	tennis match 4 pm dinner meet Anne after tennis match for pizza 6:30
7–10 pm **Study Time** Math Test, Ch 7 English: Read short stories Grammar quiz Start Spanish essay History: finish Ch 11	**Study Time** Math: pg 219 Grammar quiz finish Spanish essay Health: Ch 8 9:00 TV show	**Study Time** Math: p 221 Health: Ch 8 review quest. finish History timeline English: work on autobiography	7 pm Computer Club **Study Time** English: read Ch 8 English: finish autobiography	Football Game 7 pm Remember: Sleepover at Toni's!

Weekend: Tennis Match 10 am Sat., soccer game Sat. night, get supplies for Health project
7 - 9 pm Sun. Study Time: finish English Ch 8, start Health project

Long-Range Assignments (due next week): Health project—immune system, finish biography, English report on biography

Better Grades & Higher Test Scores / SCIENCE
Copyright ©2003 by Incentive Publications, Inc., Nashville, TN.

Get Ready: Get Organized

Date	Subject	Assignment	Due Date
4/22	English	choose biography, write report	5/16
4/23	Spanish	essay about a holiday	5/1
		3 short stories in book (pgs 235-255)	5/6
		Cold War timeline	5/8
		project on vaccines	5/5
4/30	English	Grammar quiz, Ch 7	5/1
4/30	English	autobiography	5/9
5/1	English	read Ch 4 poetry	5/9
5/1	Math	Ch 7 test	5/6
5/5	Math	even-numbered problems, page 219	5/1
5/1	Math	problems, page 221	5/8
5/5	Health	read Ch 9	5/8

Get Your Assignments Organized

You can't do a very good job on an assignment if you don't have a clue about what it is. You can't possibly do the assignment well if you don't understand the things you're studying. So, if you want to get smarter, get clear and organized about assignments. It takes 7 simple steps to do this:

1. Listen to the assignment.

2. Write it down in an assignment notebook.
   (Make sure you write down the due date.)

3. If you don't understand the assignment—ASK.
   (Do not leave the classroom without knowing what it is you are supposed to do.)

4. If you don't understand the material well enough to do the assignment—
   TALK to the teacher. (*Tell him or her that you need help getting it clear.*)

5. Take the assignment book home.

6. Transfer assignments to your weekly or monthly schedule at home.

7. Look at your assignment book every day.

Get Ready Tip # 6
Don't count on anyone else to listen to the assignment and get it down right. Get the assignment yourself. Find a reliable classmate to get assignments when you are absent—or contact the teacher directly.

Get Yourself Organized

Okay, so your schedule is on the wall—all neat and clear. Your study space is organized. Your study supplies are organized. You have written down all your assignments, and you've got all your lists made. Great! But do you feel rushed, frenzied, or hassled? Take some time to think about the behaviors that will help YOU feel as organized as your stuff and your schedule.

Before you leave school . . .

STOP—take a few calm, unrushed minutes to think about what books and supplies you will need at home for studying. ALWAYS take the assignment notebook home.

When you get home . . .

FIRST—Put your school bag in the same spot every day, out of the way of the bustle of your family's activities.

STOP—after relaxing, or after dinner, take a few calm, unrushed minutes to look over your schedule and review what needs to be done. Review your list for the day. Plan your evening study time and set priorities. Don't wait until it is late or you are very tired.

Before you go to bed . . .

STOP—take a few calm, unrushed minutes to look over the assignment notebook and the to-do list for tomorrow one more time. Make sure everything is done.

THEN—put everything you need for the next day IN the book bag. Don't wait until morning. Make sure you have all the right books and notebooks in the bag. Make sure your finished work is all in the bag. Also, pack other stuff (for gym, sports, etc.) at the same time. Put everything in one consistent place, so you don't have to rush around looking for it.

In the morning . . .

STOP—take a few calm, unrushed minutes to think and review the day one more time.

THEN—eat a good breakfast.

I'm cool! I've got everything I need.

Oh no!
Guess what Max forgot?

Max finally finished his project on the electromagnetic spectrum. He worked on it for a week. He typed it perfectly and included great illustrations and examples of creative graphs and grids. He added a great cover and a smashing title. It looks fantastic. And it's due today. He remembered to take his lunch and his gym bag. He remembered the comic books he promised to lend his friend Zack.

Get Ready Tip # 7
It doesn't do much good to get your homework done if you don't turn it in.

Get Healthy

If you are sick, or tired, or droopy, or angry, or nervous, or weak, or miserable, it is very hard to be a good student. It is hard to even use or show what you already know. Your physical and mental health is a basic MUST for doing as well as you would like to in school. So, don't ignore your health. Pay attention to how you feel. No one else can do that for you.

Get plenty of rest

If you're tired, nothing else works very well in your life. You can't think, or concentrate, pay attention, learn, remember, or study. Try to get 7 or 8 hours of sleep every night. Get plenty of rest on weekends. If you have a long evening of study ahead, take a short nap after school.

Eat well

You can't learn or function well on an empty stomach. And all that junk food (soda, sweets, chips, snacks) actually will make you more tired. Plus, it crowds the healthy foods out of your diet—the foods your brain needs to think well and your body needs to get through the day with energy. So eat a balanced diet, with lean meat, whole grains, vegetables, fruit and dairy products. Oh, and drink a lot of water—8 glasses a day is good.

Exercise

Everything in your body works better when your body gets a chance to move. Make sure your life does not get too sedentary. Do something every day to get exercise—walk, play a sport, play a game, or run. It's a good idea to get some exercise before you sit down to study, too. Exercise helps you relax, unwind, and de-stress. It's good for stimulating your brain.

Relax

Your body and your mind need rest. Do something every day to relax. Take breaks during your study time and do anything that helps you unwind.

Find relief for stress

Pay attention to signs of anxiety and stress. Are you nervous, worried, angry, sad? Are your muscles tense, your stomach in a knot? Is your head aching? Are you over-eating or have you lost your appetite? All these are signs of stress that can lower your success in school and interfere with your life. If you notice these signs, find a way to de-stress. Exercise and adequate rest are good for stress relief. You also might try these: stretch, take a hot bath, take a nice long shower, laugh, listen to calming music, write in a journal. If you're burdened with worries, anger, or problems, talk to someone—a good friend, teacher, parent, or other trusted adult.

Get a Grip (on Study Habits)

Here's some good advice for getting set to improve your study habits. Check up on yourself to see how you do with each of these. Then set goals where you need to improve.

. . . in school:

1. Go to class.

You can't learn anything in a class if you are not there. Go to all your classes. Show up on time. Take your book, your notebook, your pencil and other supplies.

2. Choose your seat wisely.

Sit where you won't be distracted. Avoid people with whom you'll be tempted to chat. Stay away from the back row. Sit where you can see and hear.

3. Pay attention.

Get everything you can out of each class. Listen. Stay awake. Your assignments will be easier if you've really been present in the class.

4. Take notes.

Write down main points. Write down examples of problems, solved correctly. If you hear it AND write it, you'll be likely to remember it.

5. Ask questions.

It's the teacher's job to see that you understand the material; it's your job to ask if you don't.

6. Use your time in class.

Get as much as possible of the next day's assignment done before you leave the class.

TOP 10 TIPS
for getting better grades

10. Get enough rest.

9. Go to class (and be on time).

8. Pay attention in class.

7. Take notes.

6. Write down your assignments.

5. Turn off the TV.

4. Don't procrastinate.

3. Do your homework.

2. Turn in your homework.

1. Ask for help before it's too late.

7. Write down assignments.

Do not leave class until you understand the assignment and have it written down clearly.

8. Turn in your homework.

If you turn in every homework assignment, you are closer to doing well in a class—even if you struggle with tests.

. . . at home:

9. Gather your supplies.

Before you sit down to study, get all the stuff together that you'll need: Assignment book, notebook, notes, textbook, study guides, paper, pencils, etc. Think ahead so that you have supplies for long-term projects. Bring those home from school or shop for those well in advance.

10. Avoid distractions.

Think of all the things that keep you from concentrating. Figure out ways to remove those from your life during study time. In other words: Make a commitment to keep your study time uninterrupted. If you listen to music while studying, choose music that can be in the background, not the foreground of your mind.

11. Turn off the TV.

No matter now much you insist otherwise, you cannot study well with the TV on. Plan your TV time before or after study time, not during.

12. Phone later.

Plan a time for phone calls. Like TV watching, phoning does not mix with focusing on studies. The best way to avoid this distraction is to study in a room with no phone. Call your friends when your work is done.

13. Hide the computer games.

Stay away from the video game playing stations, computer games, email, and Internet surfing. Plan time for these when studies are done, or before you settle into serious study time.

14. Know where you're going.

Review your weekly schedule and your assignment notebook. Be sure about what it is that needs to be done. Make a clear *To-Do* list for each day, so you will know what to study. Post notes on your wall, your refrigerator, or anywhere that will remind you what things you need to get done!

15. Plan your time.

Think about the time you have to work each night. Make a timeline for yourself. Estimate how much time a task will take, and set some deadlines. This will keep your attention from wandering and keep you focused on the task.

16. Start early.

Start early in the evening. Don't wait until 10 PM to get underway on any assignment. Whenever possible, start the day before or a few days before.

17. Do the hardest things first.

It's a good idea to do the hardest and most important tasks first. This keeps you from procrastinating on the tough assignments. Also, you'll be doing the harder stuff when your mind is the most fresh. Study for tests and do hard problems early, when your brain is fresh. Do routine tasks later in the evening.

18. Break up long assignments.

Big projects or test preparations can be overwhelming. Break each long task down into small ones. Then take one small task at a time. This will make the long assignments far less intimidating. And you'll have more successes more often. Never try to do a long assignment all in one sitting.

19. Take breaks.

Plan a break for your body and mind every 30-45 minutes. Get up, walk around, stretch, do something active or relaxing. However, avoid getting caught up in any long phone conversations or TV shows. You'll never get back to the studying!

20. Cut out the excuses.

It's perfectly normal to want to avoid doing school work. Just about everybody has a whole list of techniques for work avoidance. And the excuses people give for putting off or ignoring it are so numerous, they could fill a whole book. Excuses just take up your energy. In the time you waste convincing yourself or anyone else that you have a good reason for avoiding your studies, you could be getting some of the work done. If you want to be a better student, you'll need to dump your own list of excuses.

No More Excuses!

21. Plan ahead for long-range assignments.

Start early on long-range assignments, big projects, and test preparations. Don't wait until the night before anything is due. You never know what will happen that last day. You could be distracted, sick, or unexpectedly derailed. Get going on long tasks several days before the due date. Make a list of everything that needs to be done for a long-range assignment (including finding information and collecting supplies). Then, start from the due-date and work backwards. Make a timeline or schedule that sets a time to do each of the tasks on the list.

22. Don't get behind.

Keeping up is good. Many students slip into failure, stress, and hopelessness because they get behind. The best way to avoid all of these is simply NOT to get behind. This means DO your assignments on time.

If you do get behind because of illness or something else unavoidable, do something about it. Don't get further and further into the pit! Talk to the teacher. Make a plan for catching up.

Getting behind is often caused by procrastination. Don't procrastinate. The more you do, the worse you feel, and the harder it is to catch up!

23. Get on top of problems.

Don't let small problems develop into big ones. If you are lost in a class, missed an assignment, don't understand something, or have done poorly on something—act quickly. Talk to the teacher, ask a parent to help, find another student who has the information. Do something to correct the problem before it becomes overwhelming.

Some awesome facts turn up in science homework!

Science Fax

All the capillaries in the human body laid end to end would reach around the world twice!

The poison from the skin of one South American golden poison dart frog could kill 1500 people.

24. Ask for help.

You don't have to solve every problem alone or learn everything by yourself. Don't count on someone noticing that you need help. Tell them. Use the adults and services around you to ask for help when you need it. Remember, it is the teacher's job to teach you. Most teachers are happy to help a student who shows interest in getting help.

25. Reward yourself for accomplishments.

If you break your assignments down into manageable tasks, you'll have more successes more often. Congratulate and reward yourself for each task accomplished— by taking a break, getting some popcorn, going for a walk, bragging to someone about what you've done—or any other way you discover. Every accomplishment is worth celebrating!

GET SET →

Get Familiar with Science Tools

Formulas, measurement tools, units of measurement—these sound like math tools!
They are, but they are also science tools because math is used constantly in science.
If you're going to be successful with investigating science questions,
you'll need to have these firmly planted in your mind.

Know Your Measures

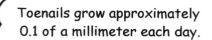

Toenails grow approximately
0.1 of a millimeter each day.

The total surface area
covered by skin
on an adult human
is about 22 square feet.

Length

Metric System

1 centimeter (cm)	=	10 millimeters (mm)
1 decimeter (dm)	=	10 centimeters (cm)
1 meter (m)	=	10 decimeters (dm)
1 meter (m)	=	100 centimeters (cm)
1 meter (m)	=	1000 millimeters (mm)
1 decameter (dkm)	=	10 meters (m)
1 hectometer (hm)	=	100 meters (m)
1 kilometer (km)	=	100 decameters (dkm)
1 kilometer (km)	=	1000 meters (m)

English System (U.S. Customary)

1 foot (ft)	=	12 inches (in)
1 yard (yd)	=	36 inches (in)
1 yard (yd)	=	3 feet (ft)
1 mile (mi)	=	5280 feet (ft)
1 mile (mi)	=	1760 yards (yd)

Area

Metric System

1 square meter (m^2)	= 100 square decimeters (dm^2)
1 square meter (m^2)	= 10,000 square centimeters (cm^2)
1 hectare (ha)	= 0.01 square kilometer (km^2)
1 hectare (ha)	= 10,000 square meters (m^2)
1 square kilometer (km^2)	= 1,000,000 square meters (m^2)
1 square kilometer (km^2)	= 100 hectares (ha)

English System (U.S. Customary)

1 square foot (ft^2)	= 144 square inches (in^2)
1 square yard (yd^2)	= 9 square feet (ft^2)
1 square yard (yd^2)	= 1296 square inches (in^2)
1 acre (a)	= 4840 square yards (yd^2)
1 acre (a)	= 43,560 square feet (ft^2)
1 square mile (mi^2)	= 640 acres (a)

Capacity

Metric System

1 teaspoon (t)	= 5 milliliters (mL)
1 tablespoon (T)	= 12.5 milliliters (mL)
1 liter (L)	= 1000 milliliters (mL)
1 liter (L)	= 1000 cubic centimeters (cm³)
1 liter (L)	= 1 cubic decimeter (dm³)
1 liter (L)	= 4 metric cups
1 kiloliter (kL)	= 1000 liters (L)

English System (U.S. Customary)

1 tablespoon (T)	= 3 teaspoons (t)
1 cup (c)	= 16 tablespoons (T)
1 cup (c)	= 8 fluid ounces (fl oz)
1 pint (pt)	= 2 cups (c)
1 pint (pt)	= 16 fluid ounces (fl oz)
1 quart (qt)	= 4 cups (c)
1 quart (qt)	= 2 pints (pt)
1 quart (qt)	= 32 fluid ounces (fl oz)
1 gallon (gal)	= 16 cups (c)
1 gallon (gal)	= 8 pints (pt)
1 gallon (gal)	= 4 quarts (qt)
1 gallon (gal)	= 128 fluid ounces (fl oz)

The average person uses 20 gallons of water a day to bathe or shower.

The human heart pumps about 40 trillion gallons of blood in an average 70-year life.

Volume

Metric System

1 cubic decimeter (dm³)	= 0.001 cubic meter (m³)
1 cubic decimeter (dm³)	= 1000 cubic centimeters (cm³)
1 cubic decimeter (dm³)	= 1 liter (L)
1 cubic meter (m³)	= 1,000,000 cubic centimeters (cm³)
1 cubic meter (m³)	= 1000 cubic decimeters (dm³)

English System (U.S. Customary)

1 cubic foot (ft³)	= 1728 cubic inches (in³)
1 cubic yard (yd³)	= 27 cubic feet (ft³)
1 cubic yard (yd³)	= 46,656 cubic inches (in³)

The average hot air balloon holds 2100 cubic meters of hot air.

The volume of Jupiter is over 367 trillion cubic miles.

Get Set: Science Tools

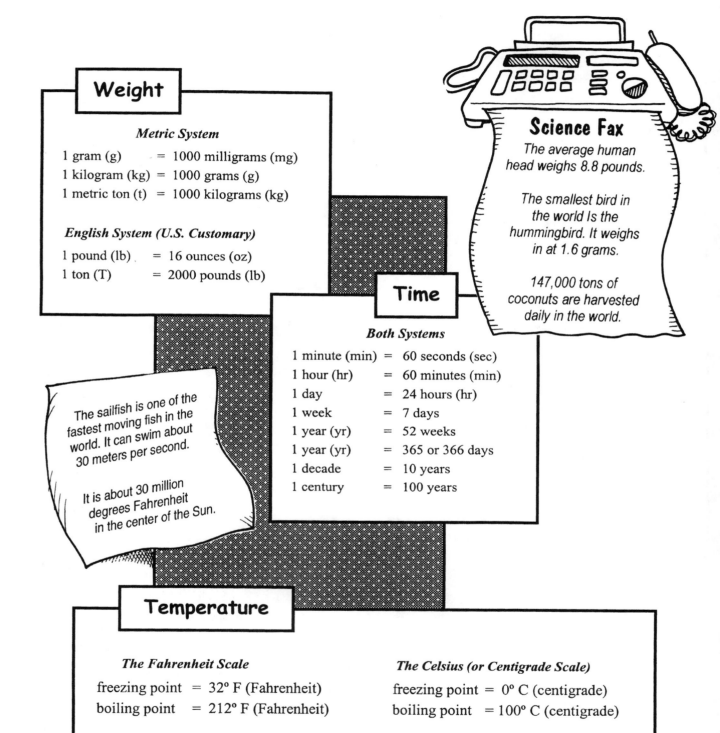

Weight

Metric System

1 gram (g)	=	1000 milligrams (mg)
1 kilogram (kg)	=	1000 grams (g)
1 metric ton (t)	=	1000 kilograms (kg)

English System (U.S. Customary)

1 pound (lb)	=	16 ounces (oz)
1 ton (T)	=	2000 pounds (lb)

Science Fax

The average human head weighs 8.8 pounds.

The smallest bird in the world Is the hummingbird. It weighs in at 1.6 grams.

147,000 tons of coconuts are harvested daily in the world.

Time

Both Systems

1 minute (min)	=	60 seconds (sec)
1 hour (hr)	=	60 minutes (min)
1 day	=	24 hours (hr)
1 week	=	7 days
1 year (yr)	=	52 weeks
1 year (yr)	=	365 or 366 days
1 decade	=	10 years
1 century	=	100 years

The sailfish is one of the fastest moving fish in the world. It can swim about 30 meters per second.

It is about 30 million degrees Fahrenheit in the center of the Sun.

Temperature

The Fahrenheit Scale

freezing point = 32° F (Fahrenheit)
boiling point = 212° F (Fahrenheit)

The Celsius (or Centigrade Scale)

freezing point = 0° C (centigrade)
boiling point = 100° C (centigrade)

Know Your Measurement Equivalents

From English to Metric

English Customary Unit	Approximate Metric Equivalent
inch	2.45 centimeters
foot	30.48 centimeters
yard	0.9144 meters
mile	1.609 kilometers
acre	4047 square meters
ounce	28.3495 grams
pound	453.59 grams
ton	907.18 kilograms
pint	0.4732 liters
quart	0.9465 liters
gallon	3.785 liters
bushel	35.2390 liters

The sound of a growling bear travels to your ears at about 1100 feet per second.

That's 335 meters per second!

From Metric to English

Metric Unit	Approximate English Equivalent
millimeter	0.04 inch
centimeter	0.39 inch
meter	39.37 inches
kilometer	3,281 feet or .62 miles
gram	0.0353 ounce
hectogram *(100 grams)*	3.53 ounces
kilogram	2.2 pounds
metric ton	22,046.6 pounds or 1.1 tons
liter	1.06 quarts

The tiny desert rat can leap 15 feet. (That's 4.5 meters!)

Temperature Conversions

To change Fahrenheit to Celsius: subtract 32, then multiply by 5/9

To change Celsius to Fahrenheit: multiply by 9/5, then add 32

Know Your Formulas

Get Set Tip # 1

Memorize these letters and symbols, so you will always know what they mean in formulas!

h = height
w = width
b = base
B = area of base
s = side
π = pi (3.14)
r = radius
d = diameter

Perimeter

$P = s + s + s$	Perimeter of a triangle
$P = 2(h + w)$	Perimeter of a rectangle
$P = \text{sum of sides}$	Perimeter of irregular polygons
$C = 2\pi r$	Perimeter or circumference of a circle
$C = \pi d$	Perimeter or circumference of a circle

Area

$A = \pi r^2$	Area of a circle
$A = s^2$	Area of a square
$A = bh$	Area of a parallelogram
$A = \frac{1}{2} bh$	Area of a triangle
$A = \frac{1}{2} (b_1 + b_2) h$	Area of a trapezoid

Volume or Capacity

$V = Bh$	Volume of a rectangular or triangular prism
$V = \frac{1}{3} Bh$	Volume of a pyramid
$V = s^3$	Volume of a cube
$V = r^2h$	Volume of a cylinder
$V = \frac{1}{3}\pi r^2 h$	Volume of a cone
$V = \frac{4}{3}\pi r^3$	Volume of a sphere

A tennis court is 36 feet wide and 78 feet long.

A tortoise crawling around the perimeter would have to travel 228 feet.

His speed is $\frac{1}{4}$ mph. If he travels without stopping, he could get around the court in about 10 minutes!

More Formulas

Speed *(velocity)* =
distance divided by time

$$V = \frac{d}{t}$$

Force =
mass times acceleration

$$F = m \times a$$

Acceleration =
initial velocity subtracted
from final velocity
divided by the time
over which
the velocity changed

$$a = \frac{V_f - V_i}{t}$$

Work =
force times distance

$$W = F \times d$$

On a safari, Aunt
Lucy is chased by
an elephant. Can
she outrun the
elephant?

An African elephant
can travel at a
velocity of 40
kilometers per hour.
This is faster than
the fastest human.

Power =
work divided by time

$$P = \frac{W}{t}$$

Electrical Energy =
the power delivered
times the length of time
it is used

$$E = P \times t$$

How much power will
the generator have to
produce for Lucy to
dry her hair and
brush her teeth in
the desert each
morning?

Electrical Current =
voltage divided by resistance

$$I = \frac{V}{R}$$

**Mechanical
Advantage** =
resistance force
divided by effort force

$$MA = \frac{F_r}{F_e}$$

Electrical Power =
voltage times current

$$P = V \times I$$

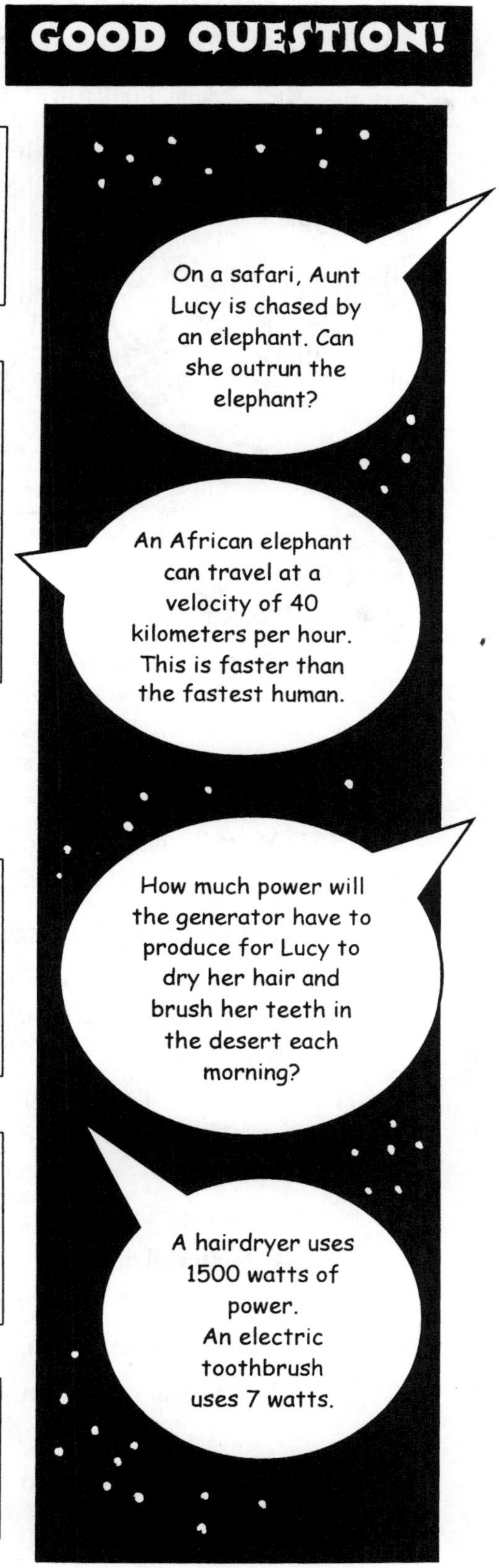

A hairdryer uses
1500 watts of
power.
An electric
toothbrush
uses 7 watts.

35

Get Tuned-Up on Thinking Skills

Aha! I think this has all the characteristics of a fungus!

Your brain is capable of an amazing variety of accomplishments! There are different levels and kinds of thinking that your brain can do—all of them necessary to get you set for good learning. To answer science questions, your brain must use many different processes. Here are some of the thinking skills that are frequently used in doing science tasks and the math tasks involved in science. Use this information to freshen up your mental flexibility and put these skills to use as you learn science concepts, use science processes, and investigate science problems.

Recall – To **recall** is to know and remember specific facts, names, processes, categories, ideas, generalizations, theories, or information.

> **Examples:** *Recall helps you remember such things as the characteristics of the three forms of matter, the names and locations of the planets, the groups and sub-groups in the system of life classification, Newton's laws of motion, the names of the bones in the human body, or how to conduct a scientific investigation.*

Classify – To **classify** is to put things into categories. When you classify ideas, numbers, topics, or things, you must choose categories that fit the purpose and clearly define each category.

> **Example:** *octopus, snail, clam, oyster, conch, squid*
>
> *There are many different ways to classify these items. They are all organisms. They are all animals. They are also all sea creatures, mollusks, and invertebrates.*

Generalize – To **generalize** is to make a broad statement about a topic based on observations or facts. A generalization should be based on plenty of evidence (facts, observations, and examples). Just one exception can prove a generalization false.

> **Examples:**
>
> **Safe Generalization:**
>> *Climates at high latitudes are likely to be cooler than those at low latitudes.*
>
> **Invalid Generalizations:**
>
>> **faulty generalization** - A faulty generalization is invalid because there are exceptions.
>>> *Average temperatures are lower at latitudes of 45° than at latitudes of 25°.*
>>
>> **broad generalization** - A broad generalization suggests something is *always* or *never* true about *all* or *none* of the members of a group. Most broad generalizations are untrue.
>>> *The climate is hot and dry at all locations between 0° and 10° latitude.*

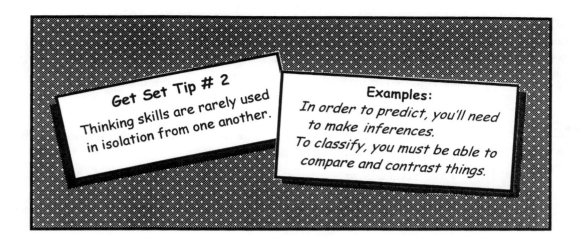

Get Set Tip # 2
Thinking skills are rarely used in isolation from one another.

Examples:
In order to predict, you'll need to make inferences.
To classify, you must be able to compare and contrast things.

Elaborate
– To **elaborate** is to provide details about a situation (to explain, compare, or give examples). When you elaborate, you might use words or phrases such as: *so, because, however, but, an example of this is, on the other hand, as a result, in addition, moreover, for instance, such as, if you recall, furthermore, another reason is.*

> **Example:** *Molecules move farther apart when a substance is heated. For instance, the molecules in ice cubes will move apart as the ice sits in the sun. This will cause the ice to turn into a liquid.*

Predict
– To **predict** is to make a statement about what will happen. Predictions are based on some previous knowledge, experience, or understanding.

> **Example:** *The trees on the hill behind Axel's home burned last summer. All the underground brush also burned. This left bare, dusty ground. Axel predicts that there will be erosion and mudslides when the rains come this fall.*

I predict a huge mess when the rains come!

Infer
– To **infer** is to make a logical guess based on information.

> **Example:** *The plants out in Ann's yard were dead on a cold morning. The plants close to the house looked healthy. She inferred that frost had settled on the plants in the yard.*

Recognize Cause and Effect
– When one event occurs as the result of another event, there is a **cause-effect relationship** between the two. Recognizing causes and effects takes skill. When reading a science equation or problem, pay careful attention to words or symbols that give clues to cause and effect (*the reason was, because, as a result, consequently, so*).

> **Example:** *The hinged leaves of the Venus's fly-trap snapped shut* (effect) *because an insect landed on them.* (cause)

37

Hypothesize

Hypothesize – To **hypothesize** is to make an educated guess about a cause or effect. A hypothesis is based on examples that support it but do not prove it. A hypothesis is something that can be—and should be—tested.

> **Example:** *If salt is added to water in a glass, an egg will float in the water.*

Extend

Extend – To **extend** is to connect ideas or things together, or to relate one thing to something different, or to apply one idea or understanding to another situation.

> **Example:** *You learn that a crab is a crustacean because it has a segmented body with two main regions, a hard exoskeleton, two pairs of antennae, and claw-like legs at the anterior end. Then you decide that a lobster (which has the same characteristics) is also a crustacean.*

Compare & Contrast

When you **compare** things, you describe similarities.
For instance, you're short and I'm short.
I have two feet and you have two feet. You live in space and I live in space.
I am smart and you are smart.

When you **contrast** things, you describe the differences. For instance, You have two eyes and I have one eye. You have two arms and I have several arms. You are ordinary and I am extraordinary!

Draw Conclusions

Draw Conclusions – A **conclusion** is a general statement that someone makes after analyzing examples and details. A conclusion generally involves an explanation someone has developed through reasoning.

> **Example:** *Maxie bangs on bottles with various amounts of liquid in them. She notices that there are different pitches from the different bottles. She notices that the bottle with 1 inch of liquid gives a lower pitch than the bottle with about 5 inches of liquid.*
>
> *She draws these conclusions:*
>
> *1) The amount of air in a bottle affects the sound made when banging on the bottle.*
>
> *2) The greater the amount of air in the bottle, the lower the pitch which results from tapping.*

38

Analyze - To **analyze,** you must break something down into parts and determine how the parts are related to each other and how they are related to the whole.

> **Examples:** *You must analyze to . . .*
>
> *. . . identify the different organs in the digestive system and describe their functions.*
>
> *. . . describe the numbers and kinds of different atoms in a sodium chloride molecule.*
>
> *. . . explain the difference between characteristics of an insect and an arachnid.*
>
> *. . . discuss the functions of different organisms in an ecosystem.*

Synthesize - To **synthesize,** you must combine ideas or elements to create a whole.

> **Examples:** *You must synthesize to . . .*
>
> *. . . understand how muscles, tendons, ligaments, and bones work together to move the body.*
>
> *. . . explain how the movement of individual water molecules creates an ocean current.*
>
> *. . . understand how salt and warm water mix together to form a solution.*
>
> *. . . create a graph to show the results of an experiment.*

Think Logically (or Reason) – When you think **logically,** you take a statement or situation apart and examine the relationships of parts to one another. You reason **inductively** *(start from a general principle and make inferences about the details)* or **deductively** *(start from a group of details and draw a broad conclusion or make a generalization).*

> **Example:** *An insect has six legs and three body segments.*
> *The creature that bit Chester's toe had eight legs.*
> *The biting creature could not have been an insect.*

I deduce that this creature is not an insect.

Evaluate - To **evaluate** is to make a judgment about something. Evaluations should be based on evidence. Evaluations include opinions, but these opinions should be supported or explained by examples, experiences, observations, and other forms of evidence.

> **Examples:** *When you evaluate an argument, an explanation, a decision, a prediction, an inference, a conclusion, or a generalization, ask questions like these:*
>
> > *Are the conclusions reached based on good examples and facts?*
> > *Is there good evidence for the generalization or inference?*
> > *Is this believable?*
> > *Does the explanation make sense?*
> > *Are the sources used to make the decisions reliable?*
> > *Is the argument effective?*
> > *Is it realistic?*

Get Serious about Study Skills

Better Listening

Keep your ears wide open! You can increase your understanding of science concepts and processes if you listen well. Here are some tips for smart listening. They can help you get involved with the information instead of letting it just buzz by your ears.

1. Realize that the information is important.

Here's what you can get when you listen to someone who is talking to you about science:

. . . details about how a science process works
. . . help answering science questions
. . . examples of problems solved correctly or investigations done correctly
. . . hazards or difficulties you might face when doing a science process
. . . new, exciting information about how something in the world works
. . . meanings of terms used in science questions or assignments
. . . directions for certain assignments

2. Be aware of the obstacles to good listening.

Know ahead of time that these will interfere with your ability to listen well. Try to avoid them, alter them, or manage them, so they don't get in the way.

. . . fatigue   . . . wandering attention
. . . surrounding noise   . . . too many things to hear at once
. . . uncomfortable setting   . . . missing the beginning or ending
. . . personal concerns, thoughts, or worries   . . . talking

3. Make a commitment to improve.

You can't always control all obstacles (such as the comfort of the setting or the quality of the speaker's presentation), but there are things you can control. Put these to work to gain more from your listening.

. . . Get enough rest.
. . . Do your best to be comfortable while you listen.
. . . Cut out distractions. Keep your mind focused on what is being said.
. . . Look directly at the speaker.
. . . Take notes. Write down sample problems the speaker shows or solves.
. . . As the speaker talks, think of examples or relate the information to your life.
. . . Pay special attention to opening and closing remarks.
. . . Pay special attention to anything that is repeated.
. . . As soon as possible after listening, summarize or review what you have heard.

40

Careful Reading

There is plenty of reading in science. Textbooks and other materials explain science concepts and processes. Many science problems are more than just facts or numbers—they include ideas that you need to interpret. All science questions or assignments include some sort of instructions to follow. So, to succeed in science, you need to make good use of reading skills.

Before you read a science assignment or problem, have a clear idea of the purpose for reading. Are you reading to find directions for the assignment? Are you reading to learn how to do a science process? Are you reading to solve a problem? Are you reading to find a particular fact? In all these cases above, you need to read closely and carefully. To learn or review a process, or to solve a problem, you will need to read the information through more than once.

Sample Assignment:

Read the characteristics of echinoderms and the description of each organism. Circle any names that answer the question.

Echinoderm is the name given to one phylum of animals that live in salt water. They have radial symmetry and a tough outer covering of stiff spines. They also have rows of tube feet on the undersides of their bodies. These feet act like suction cups to help the animal move and grab onto food.

Lucy examined several animal species. She noticed that the octopus had bilateral symmetry. The snail she observed could live on land. The starfish had tube feet and a spiny covering. She noticed that the clam had a thick muscular foot and a hard shell covering. The sponge had an irregular shape with no symmetry. The sea urchin and the sand dollar both showed radial symmetry and had spiny coverings. Which of the many organisms that Lucy examined could not possibly be echinoderms?

If you read the problem carefully, you will . . .

. . . find clear directions for the assignment.

Read the characteristics of echinoderms.
Read the description of the organisms.
Circle any names that answer the question.

. . . identify the problem that needs to be solved or question that must be answered.

Which organisms could not possibly be echinoderms?

. . . find the information needed to help you decide how to answer the question.

Echinoderms live in salt water.
Echinoderms have radial symmetry.
Echinoderms have a tough outer covering.
Echinoderms have stiff spines.
Echinoderms have tube-like feet.
The octopus had bilateral symmetry.
The snail could live on land.
The starfish had tube feet and a spiny covering.
The clam had a thick muscular foot.
The clam had a hard shell covering.
The sponge had no symmetry.
The sea urchin had radial symmetry and spines.
The sand dollar had radial symmetry and spines.

Better Grades & Higher Test Scores / SCIENCE
Copyright ©2003 by Incentive Publications, Inc., Nashville, TN.

Get Set: Study Skills

Taking Notes

A study skill of major importance is knowing how to take notes well and use them effectively. Good notes from classes and from reading are valuable resources to anyone who's trying to do well as a student. A lot of learning goes on while you're taking notes—you may not even realize it's happening!

I'd better write this down.

1. **Listen or read for a main idea, key formula, or important process. Write it down.**

I've got a description of each group of rocks.

2. **Write examples, problems, definitions, and details to support each key idea.**

I think I can remember this now.

3. **Review your notes as soon as possible after the class. This will help to fix the information in your brain.**

When you take notes

1 . . . you naturally listen better. (You have to listen in order to get the information and write it down!)

2 . . . you listen differently—you naturally learn and understand the material better. Taking notes forces you to focus on what's being said or read.

3 . . . you sort through the information and decide what to write. This means you naturally think about the material and process it—making it more likely that you'll remember it.

4 . . . the actual act of writing the notes fixes the information more firmly in your brain.

5 . . . you end up with good notes in your notebook. Having written examples of the characteristics, facts, formulas, or science processes makes it possible to see the correct information. This makes it much easier to review and remember it.

Get Set Tip # 4
Take good notes! You'll remember the material better with less studying.

Tips for Wise Note-Taking

in class . . .

- Have a notebook or a notebook section for science.
- When the class begins, write the topic for the day at the top of a clean page.
- Write the date at the top of the page.
- Only take notes on one side of the paper.
- Use an erasable pen for clear notes, not a pencil.
- Write down examples of situations or problems.
- Write notes to yourself about explanations for different events or situations.
- Write neatly so you can read it later.
- Leave sizable margins to the left of the outline.
- Use this space to note important items or write key words.
- Leave a blank space after each main idea section.
- Pay close attention to the opening and closing remarks.
- Listen more than you write.
- ASK about anything you do not understand.

Get Set Tip # 5

When you take notes in class, be alert for signals from the teacher about important ideas.
Write down anything the speaker (or teacher)...
...writes on the board.
...gives as a definition.
...emphasizes with his or her voice.
...repeats.
...says is important.

. . . from a textbook assignment

- Skim through one section at a time to get the general idea. (Use the textbook divisions as a guide to separate sections, or read a few paragraphs at a time.)

- Then go back and write down the main ideas.

- For each main idea write a few supporting details or examples. If a science operation or process is explained, write down a correct example.

- Notice bold or emphasized words or phrases. Write these down with definitions.

- Read captions under pictures. Pay attention to facts, tables, charts, graphs, and pictures, and the explanations that go along with them. Put information in your notes if it is very important.

- Don't write too little. You won't have all the main points or enough examples.

- Don't write too much. You won't have time or interest in reviewing the notes.

Get Set: Study Skills

How to Prepare for a Test

Good test preparation does not begin the night before the test.
The time to get ready for a test starts long before this night.
Here are some tips to help you get ready – weeks before the test and right up to test time.

1. Start your test preparation at the beginning of the year—
or at least as soon as the material is first taught in the class.

The purpose of a test is to give a picture of what you are learning in the class. That learning doesn't start 12 hours before the test. It starts when you start attending the class. Think of test preparation this way, and you'll be able to be less overwhelmed or anxious about an upcoming test.

You'll be much better prepared for a test *(even one that is several days or weeks away)* if you . . .

. . . *pay attention in class.*

. . . *take good notes and work out sample problems.*

. . . *keep your notes and class handouts organized.*

. . . *read all your assignments.*

. . . *keep a good list of key events and definitions.*

. . . *do your homework regularly.*

. . . *make up any work you miss when you're absent.*

. . . *ask questions in class about anything you don't understand.*

. . . *review notes and handouts regularly.*

2. Once you know the date of the test, make a study plan.

Look over your schedule and plan time to start organizing and reviewing material.
Allow plenty of time to go through all the material.
Your brain will retain more if you review it a few times and spread the studying out over several days.

3. Get all the information you can about the test.

Write down everything the teacher says about the test.
Get clear about what material will be covered.
If you can, find out about the format of the test.
Make sure you get all study guides the teacher distributes.
Make sure you listen well to any in-class reviews.

Get Set: Study Skills

Better Grades & Higher Test Scores / SCIENCE
Copyright ©2003 by Incentive Publications, Inc., Nashville, TN.

4. Use your study time effectively.

DOs and DON'Ts

- **DO** gather and organize all your notes and handouts.
- **DO** review your text; pay attention to bold words, bold statements, and examples of concepts, processes, or questions.
- **DO** identify the kinds of problems in the section being tested; practice solving a few of each kind.
- **DO** review the questions at the end of text sections; practice answering them.
- **DO** review your notes, using a highlighter to emphasize important points.
- **DO** review the study guides provided by the teacher.
- **DO** review any previous quizzes on the same material.
- **DO** predict the questions that may be asked and kinds of problems that will be included; think about how you would answer them.
- **DO** make study guides and aids for yourself.
- **DO** make sets of cards with key vocabulary words, terms and definitions, main concepts, formulas, or facts.
- **DO** ask someone (reliable) to quiz you on the main points and terms.

- **DON'T** spend your study time blankly staring your notebook or or mindlessly leafing through your textbook.
- **DON'T** study with someone else unless that person actually helps you learn the material better.
- **DON'T** study when you're hungry or tired.
- **DON'T** study so long at one time that you get tired, bored, or distracted.

5. Get yourself and your supplies ready.

Do these things the night before the test (not too late):

Gather all the supplies you need for taking the test (good pencils with erasers, erasable pens, scratch paper, calculator with batteries).

Put these supplies in your school bag.

Gather your study guides, notes, and text into your school bag.

Get a good night of rest.

In the morning:

Eat a healthy breakfast.

Look over your study guides and note card reminders.

Relax and be confident that your preparation will pay off.

How to Take a Test

Before the test begins

- Get to class on time, or even a bit early, so you don't have to rush or feel extra stressed.

- Have supplies ready. Take sharpened pencils, scratch paper, calculator, and eraser.

- Try to get a little exercise before class to help you relax.

- Go to the bathroom and get a drink.

- Get settled into your seat; get your supplies out.

- If there's time, you might glance over your study guides while you wait.

- To relax, take some deep breaths and exhale slowly.

When you get the test

- Put your name on all pages.

- Before you write anything, scan over the test to see how long it is, what kinds of questions it has, and generally what it includes.

- Think about your time and quickly plan how much time you can spend on each section.

- Read each set of directions twice. Circle key words in the directions.

- Answer all the short-answer questions. Do not leave any blanks.

- If you are not sure of an answer, make a smart guess.

- Don't change an answer unless you are absolutely sure it is wrong.

Lucy is not sure of the answer. So she makes a smart guess. She puts an **X** by the problem so she will remember to come back to it later.

When she comes back to the question, she is still not sure, so she stays with her first answer.

Get Set Tip # 7

Research shows that your first answer is correct more often than not! So stick with it unless you are positive about another answer.

— More Test-Taking Tips —

Tips for Answering Multiple Choice Questions

Multiple choice questions give you several answers from which to choose.

- Read the question through twice.
- Before you look at the choices, close your eyes and answer the question. Then look for that answer.
- Read all the choices through before you circle one.
- If you are not absolutely sure, cross out answers that are obviously incorrect.
- Choose the answer that is most complete or most accurate.
- If you're not absolutely sure, choose an answer that has not been ruled out.
- Do not change an answer unless you are absolutely sure of the correct answer.

Tips for Answering Matching Questions

Matching questions ask you to recognize facts or definitions in one column that match facts, definitions, answers, or descriptions in a second column.

- Read through both columns to familiarize yourself with the choices.
- Do the easy matches first.
- Cross off answers as you use them.
- Match the left-over items last.
- If you don't know the answer, make a smart guess.

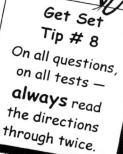

Get Set
Tip # 8
On all questions, on all tests — **always** read the directions through twice.

Tips for Answering Fill-in-the Blank Questions

Fill-in-the-blank questions ask you to write a word that completes the sentence.

- Read through each question. Answer it the best you can.
- If you don't know an answer, **X** the question and go on to do the ones you know.
- Go back to the **X**'d questions. If you don't know the exact answer, write a similar word or definition—come as close as you can.
- If you have no idea of the answer, make a smart guess.

Tips for Answering True-False Questions

True-False questions ask you to tell whether a statement is true or false.

- Watch for words like *most, some,* and *often.* Usually statements with these words are TRUE.
- Watch for words like *all, always, only, none, nobody,* and *never.* Usually statements with these words are FALSE.
- If any part of a statement is false, then the item is FALSE.

Even More Test-Taking Tips

Tips for Solving Word Problems Involving Math

A **word problem** uses words to describe a problem or question which needs a solution.

- First, read through the problem twice.
- Identify the question to be answered. Underline it.
- Circle key facts needed to solve the problem.
- Circle clue words that point to the correct operation.
- Choose a strategy for solving the problem.
- Write down a problem or equation that could solve the problem.
- Draw diagrams, charts, or pictures if you need them.
- Solve the problem.
- Go back and read the problem again. Ask yourself:

    Did I answer the question that the problem required?
    Is this answer reasonable?

- Check your answer using another method or strategy, if you have time.

Tips for Solving Number Sentences or Equations

A **number sentence or equation** is a problem made up of numbers, usually with a missing number for you to find. In an equation, the missing number is often represented by a letter.

- Read the number sentence twice.
- Identify the missing element or number that you need to find.
- Simplify the sentence or equation by combining elements that are the same, or by doing easy computations.
- Identify the operation needed to solve the problem.
- Solve the problem.
- Take your answer and write it into the number sentence or equation.
- Read through the problem again to make sure it is correct with your answer inserted.

GET SHARP →

on Science Concepts & Processes

The Nature of Science

The Science Times

The Newspaper of Science History & Ideas

Science is a way of learning about the natural world.

Scientists study many different substances and topics, but they all explore the way the world works. This way of learning called *science* has some special characteristics to know.

Science Has Existed Throughout History and Across Cultures

The Wheel is Over 5500 Years Old

Almost six thousand years ago, a simple wooden wheel changed transportation forever. Now wheels made of many different substances travel at speeds up to hundreds of miles per hour.

- Scientific discoveries have been taking place for as long as human life has existed.
- Important contributions to science have been made by men and women of many different cultures, countries, nationalities, ethnic backgrounds, ages, and beliefs.
- Scientists have a wide range of interests, skills, professions, and abilities.
- Scientists work in many different places and ways, often in cooperation with other scientists.
- You don't have to be a professional scientist to think scientifically or to investigate a question with a scientific approach.

Secretary's Invention Solves Chronic Problem

In the 1950s, a single working mother had a great idea. Like other secretaries, Bette Nesbith Graham was plagued by messy erasures of typed mistakes. So she created a brush-on correction fluid, now known as *Liquid Paper.*

The Science-Society Connection

- Science affects the daily lives of humans in thousands of ways.
- Scientific advances have long-lasting effects on people, society, and the planet.
- Scientific research and the uses of science are affected by the needs, values, and politics of a society.
- Science in itself is not good or bad, but uses of science may have benefits or harmful consequences (or both).

News Flash!
Congress debates the ethics and societal impact of stem-cell research.

Science Is a Human Endeavor

- Science is done by humans. This means it includes human motivations, interpretations, opinions, and biases.
- Good scientific research requires many human qualities such as curiosity, creativity, honesty, and judgment.
- Scientific investigations and conclusions are influenced by personal, social, and cultural beliefs and values.

WOW!

Science Involves a Particular Way of Knowing Things

- Scientists, students, and other curious persons do investigations to find answers to questions about the way the world works.
- Scientific investigation involves observation, imagination, and logical thinking.
- Scientists use systematic methods to collect facts and observations. Then they use these to build theories that explain how or why things happen.
- Science places heavy emphasis on evidence.
- A theory is accepted after someone else tests it. To become a part of scientific knowledge, the same results must be found from repeated tests.
- Scientists publish the methods and conclusions of investigations so they can be reviewed by others. (Review by an outside party is called *peer review*.)
- Scientists researching the same question or problem may find and publish different results.
- Skepticism is valued in science. This means that any theory or result is open to questions, criticism, and close scrutiny.
- Many scientific conclusions and explanations raise more questions.

Do Cell Phones Cause Cancer?

Concerned citizens and consumer advocates wonder about the safety of cellular phones. This question has not yet been fully answered by scientific research.

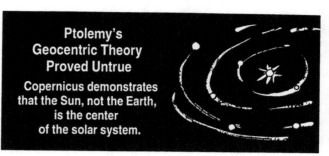

Ptolemy's Geocentric Theory Proved Untrue

Copernicus demonstrates that the Sun, not the Earth, is the center of the solar system.

Scientific Knowledge Is Constantly Changing

- All scientific ideas, laws, theories, and knowledge are subject to change and growth.
- Scientific knowledge is always building on earlier knowledge.
- Scientific knowledge changes as new evidence is formed. Old theories are revised, and new theories develop.

NEW ELEMENT DISCOVERED

The Periodic Table will need to be revised again after the production of element 110 in 1994. By early 2002, the element had not yet been named.

SCIENCE HAS LIMITS!

- Science can only answer some questions or solve some problems. There are many human problems that science cannot solve. Scientific advances can actually create some problems.

BREAKING NEWS!

New bacteria is resistant to all known antibiotics.

Scientists Clone Sheep!

Science & Technology

Science is a major part of our everyday lives—mostly because of the technology that science has made possible. Modern technology affects just about every part of our lives—the way we travel, move, eat, play, communicate, learn, and work.

• Scientific experiments and discoveries lead to the invention and creation of technological tools.

For example: The discovery of electricity made many machines and instruments possible (such as the electromagnet, the vacuum cleaner, the light bulb).

• On the other hand, technological tools or instruments are needed to help scientists make discoveries and answer questions.

For example: The telescope made it possible for astronomers to discover planets. Powerful microscopes made it possible for scientists to learn about the existence of atomic particles.

• Besides being ever-present in our personal lives, science and technology play roles in many issues outside our homes—local, national, and global.

• The choices that people (or societies) make about how to use science have long-lasting effects on individuals, natural resources, and ecosystems. *(Think about choices such as whether to do widespread cutting of trees in the rainforest, how much control a government should have over the Internet, or how to balance all the needs for water in a county or other region.)*

• Most instruments of technology can provide benefits for human life. At the same time, these technologies have costs or negative consequences. *(Think about the benefits and costs of some technologies such as fast racing skis, sound amplifiers, DVD players, jet engines, microwave ovens, video games, computers, gunpowder, roller coasters, or chain saws.)*

• Humans must constantly make judgments and ethical decisions about how to put science to use.

Here are just a few examples of ways a technology can spell good news or bad news for humans, or for the planet Earth.

Good News: The Internet
- instant communication
- easy, quick information, news, music, and entertainment
- easy shopping
- a great timesaver

Bad News: The Internet
- thousands of questionable sites
- pornography easily available
- plenty of unreliable information
- surfing the Web eats up time
- time on Web isolates people from live contact

Good News: Automatic Saws
- far fewer mill employees needed
- greater level of safety
- owners save on wage costs
- saws are fast (increased production)

Bad News: Automatic Saws
- job loss for workers
- saws cost money, may increase lumber prices
- use up natural resources for power

Good News: Dams on Mississippi
- keep the river from flooding and damaging property

Bad News: Dams on Mississippi
- interfere with the natural processes of river flow and river deposits
- make cities and land vulnerable to flooding from a tide surge caused by a hurricane

Good News: Germs Produced in Lab
- scientists can create vaccines to protect against disease

Bad News: Germs Produced in Lab
- labs can create germs for biological warfare

Good News: Credit Cards
- convenience
- don't have to carry cash
- fuels the economy
- buy now, pay later

Bad News: Credit Cards
- people get into debt over their heads
- large interest payments
- people buy more than they need
- buy now, pay later

Good News: City Bus System
- convenient way to get around
- reduces pollution from cars
- cheaper than owning a car
- safer than walking in some areas

Bad News: City Bus System
- noise pollution
- exhaust pollution
- danger to pedestrians
- keeps people from healthy walking

Get Sharp: Science & Technology

Theories & Laws

Some Key Theories

Big Bang Theory – The universe formed as the result of a giant, violent explosion.

Cell Theory – The cell is the basic structural and functional unit of all plants and animals.

Chaos Theory – Systems behave unpredictably and randomly even though they clearly appear to be governed by well-understood laws of physics.

Continental Drift Theory – The continents were once a single land mass, but have moved from their original locations.

Electromagnetic Theory – Electric and magnetic fields act together to produce electromagnetic waves of radiant energy.

Germ Theory – Infectious diseases are caused by microorganisms.

Heliocentric Theory – Earth and the other planets revolve around the Sun.

Plate Tectonics Theory – The Earth has an outer shell of rigid plates that move about on a layer of hot, flowing rock.

Quark Theory – The nuclear material of atoms is made up of subatomic particles.

Theory of Evolution – All species of plant and animal life developed gradually from a small number of common ancestors.

Theory of Relativity – Observations of time and space are relative to the observer.

Theory of Superconductivity – The electrical resistance of a substance disappears at very low temperatures.

Some Key Scientific Laws & Principles

Archimedes' Principle – The loss of weight of an object in water is equal to the weight of the displaced water.

Beer's Law – No substance is perfectly transparent, but some of the light passing through the substance is always absorbed.

Beodes' Law – This law is an empirical rule that gives the approximate relative distances of the planets from the Sun.

Bernoulli's Principle – The pressure of a fluid increases as its velocity decreases, and decreases as its velocity increases.

Boyle's Law – The volume of a fixed amount of gas varies directly with the pressure of the gas, provided that there is no change in temperature.

Charles' Law – The volume of a fixed amount of gas varies directly with the temperature of the gas, provided the pressure does not change.

Law of Conservation of Matter – Matter is neither created nor destroyed in a chemical change, but is only rearranged.

Law of Hydrostatics – The pressure caused by the weight of a column of fluid is determined by the height of the column.

Law of Universal Gravitation – A gravitational force is present between any two objects. The size of the force depends on the masses of the two objects and the distance between the two objects.

Mendel's Laws – Units called genes, which occur in pairs, determine heredity characteristics.

Newton's First Law of Motion (Law of Inertia) – A mass moving at a constant velocity tends to continue moving at that velocity unless acted upon by an outside force.

Newton's Second Law of Motion (Law of Action) – The acceleration of an object depends upon its mass and the applied force.

Newton's Third Law of Motion (Law of Reaction) – For every action there is an equal and opposite reaction.

Newton's Law of Gravitation – All objects exert an attractive force on one another.

Ohm's Law – Electromagnetic force equals the electric current multiplied by the resistance in a circuit.

Pascal's Law – Pressure that is applied to a fluid enclosed in a container is transmitted with equal force throughout the container.

Principle of Uniformitarianism – The processes that act on Earth's surface today are the same as the processes that have acted upon Earth's surface in the past.

Different Kinds of Science

Branches of Science

Scientific study is generally divided into major groups, called branches.
Many smaller categories of study (fields) exist within each branch.

My field is astrophysics.

It's part physical science and part Earth-space science.

The Physical Sciences focus on the structure, properties, and behaviors of matter (non-living).

The Life Sciences focus on the structure and function of living organisms and their interaction with the physical environment.

The Earth & Space Sciences focus on the composition, history, forces, and movements of Earth, the solar system, and the universe.

The Social Sciences focus on the people, groups, and institutions of human society, examining the relationships between individuals and the social groups to which they belong.

Mathematics & Logic focus on reasoning and processes which need numbers (measurement, calculations, predictions).

Fields of Science

What is studied within each of these fields?

aerodynamics – mechanics of motion between air and a solid in air
agronomy – soil and crop-raising
algebra – solving equations
anatomy – parts of organisms and their relationships to each other
anthropology – human cultures
archeology – past human cultures and cultural development, and the items made and used by the cultures
arithmetic – numbers and methods for calculating with numbers
astronomy – objects in space and the history and structure of the universe
astrophysics – physical processes in the workings of objects in space

atomic physics – structure and properties of atoms
bacteriology – bacteria
biochemistry – chemical makeup of living things and life processes
biology – living things
biophysics – physical processes in the functions of living things
botany – plants
calculus – solving problems with changing quantities
cardiology – heart function and diseases
chemistry – the composition, structure, and changes of substances
climatology – Earth's climates (weather patterns and trends)

cryogenics – behavior of matter at very low temperatures

cytology – structures of cells

ecology – relationships of living things to each other and to their environment

economics – systems by which people produce, distribute, and use goods and services

electronics – application of scientific technology of electricity

embryology – fetal development

entomology – insects and insect control

genetics – heredity and genes

geochronology – age and history of the Earth and its parts

geography – physical structures of Earth's surface and their relationship to human life and cultures

geology – the history, composition, and structure of the Earth

geometry – relationship of points, lines, surfaces, figures, and solids in space

herpetology – reptiles

histology – tissues

ichthyology – fish

immunology – immune system

molecular biology – structure and functions of particular essential molecules such as proteins

nuclear physics – structure, composition, and behavior of the nuclei of atoms

oceanography – physical properties, processes, and inhabitants of the ocean

oncology – cancer

organic chemistry – compounds containing the element carbon

ornithology – birds

paleontology – forms of life from prehistoric times

pathology – diseases

marine biology – all life in the sea

medicine – study, prevention, and treatment of disease, injury, and sickness

meteorology – Earth's atmosphere and weather

microbiology – organisms that can only be seen with the aid of a microscope

mineralogy – minerals

petrology – rocks

physics – matter and energy

political science – politics, power, and governments

probability – the likelihood that an event will occur

psychology – human behavior

physiology – functions of living things and their parts

radiology – X-rays and other forms of radiant energy

seismology – earthquakes

sociology – interrelationships of people in human societies and communities

statistics – analysis of large amounts of numerical data

taxonomy – classification of living things

thermodynamics – heat as produced by the motion of molecules

zoology – animals

57

Some Dazzling Events

These are just a few of the remarkable, fascinating, life-changing discoveries, inventions, and events that mark the history of science. You can do further research to learn more about any of these.

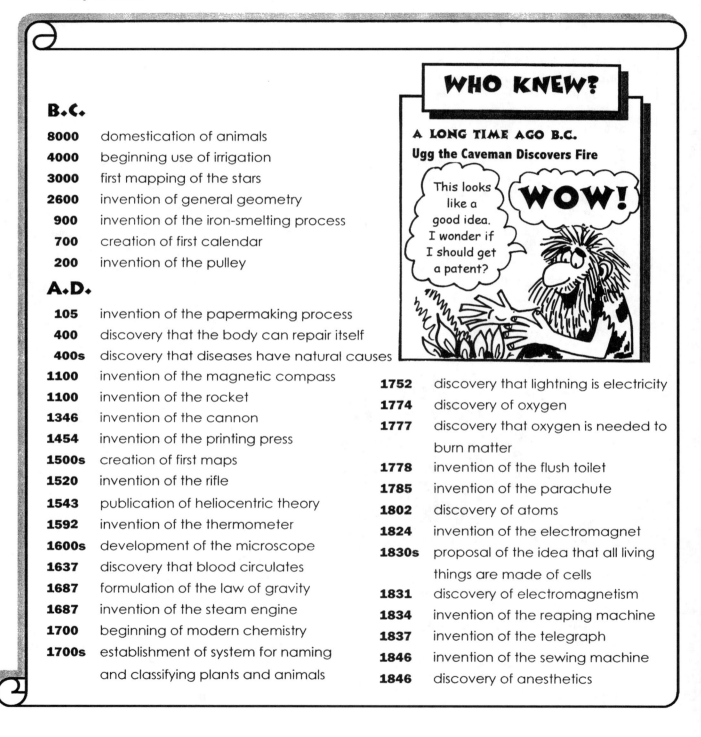

B.C.

8000	domestication of animals
4000	beginning use of irrigation
3000	first mapping of the stars
2600	invention of general geometry
900	invention of the iron-smelting process
700	creation of first calendar
200	invention of the pulley

A.D.

105	invention of the papermaking process
400	discovery that the body can repair itself
400s	discovery that diseases have natural causes
1100	invention of the magnetic compass
1100	invention of the rocket
1346	invention of the cannon
1454	invention of the printing press
1500s	creation of first maps
1520	invention of the rifle
1543	publication of heliocentric theory
1592	invention of the thermometer
1600s	development of the microscope
1637	discovery that blood circulates
1687	formulation of the law of gravity
1687	invention of the steam engine
1700	beginning of modern chemistry
1700s	establishment of system for naming and classifying plants and animals

WHO KNEW?

A LONG TIME AGO B.C.
Ugg the Caveman Discovers Fire

This looks like a good idea. I wonder if I should get a patent?

WOW!

1752	discovery that lightning is electricity
1774	discovery of oxygen
1777	discovery that oxygen is needed to burn matter
1778	invention of the flush toilet
1785	invention of the parachute
1802	discovery of atoms
1824	invention of the electromagnet
1830s	proposal of the idea that all living things are made of cells
1831	discovery of electromagnetism
1834	invention of the reaping machine
1837	invention of the telegraph
1846	invention of the sewing machine
1846	discovery of anesthetics

1849	invention of the safety pin	**1939**	invention of the helicopter
1852	invention of the passenger elevator	**1940**	invention of the photocopier
1858	invention of the Bunsen burner	**1947**	the first breaking of sound barrier
1858	invention of the electric refrigerator	**1953**	building a model of DNA
1859	publication of *Theories of Evolution*	**1953**	broadcast of the first color TV show
1866	invention of dynamite	**1954**	building of first nuclear submarine
1867	invention of the typewriter	**1956**	first videotape recording
1869	publication of first periodic table	**1957**	launching of *Sputnik*, the first
1870	invention of margarine		human-made satellite to orbit Earth
1876	invention of the telephone	**1959**	invention of the microchip
1877	invention of the sound recording	**1960**	invention of the laser
1879	invention of electric light bulb	**1965**	invention of the holograph
1879	invention of the cash register	**1969**	broadcast of the first moon walk
1880s	discovery of the laws of heredity	**1971**	invention of the microprocessor
1880s	discovery of the causes of diseases	**1972**	invention of the CD
1885	invention of the motorcycle	**1981**	launch of the first space shuttle
1892	invention of the zipper	**1983**	isolation of the virus that causes AIDS
1895	invention of the diesel engine	**1990**	launching of Hubble Space Telescope
1895	demonstration of the first radio	**1991**	arrival of World Wide Web;
1895	discovery of X-rays		beginning of widespread Internet use
1898	invention of the submarine	**1995**	first full operation of the GPS
1898	discovery of radium	**2000**	mapping of the human genome
1898	invention of the camera	**2001**	discovery of a large celestial body in
1899	invention of the tape recorder		the solar system beyond Pluto *(Quaoar)*
1901	invention of the washing machine		
1902	invention of the air conditioner		
1903	first flight of engine-powered airplane		**A.D. 2002**
1905	publication of *Theory of Relativity*		Another discovery . . .
1910	invention of neon lights		
1925	invention of the television		
1929	discovery of penicillin		
1930s	discovery of atomic energy		
1930	invention of the jet engine		
1930	discovery of Pluto, the 9th planet		Is that a speck on
1935	invention of nylon *(first synthetic fiber)*		my lens, or have I
1938	invention of the ballpoint pen		just discovered a
			new planet?

59

Some Memorable Scientists

These are just a few of the hard-working, creative people who, over the years have made some astounding discoveries and accomplished some important work. In most cases, these people did a lifetime of work searching for answers. You can explore their lives and work more completely. Find out more about their work! Find out what other contributions they made to science.

Science Fax:

It is a little known fact that the actress Hedy Lamarr co-invented an important technology for radio communications called *frequency hopping* during World War II. This technology is now called *speed spectrum*, and is the scientific basis for cordless and wireless phones and many other innovations. If she had renewed her patent when it expired, Hedy would have been a rich woman. Unfortunately, she died practically penniless.

The holes in this player piano music give me an idea for multiple radio frequencies.

Aristotle (300s B.C.) – noted for his works on logic, metaphysics, ethics, and politics

Sara Josephine Baker (early 1900s) – invented safe infant clothing; found a way to make silver nitrate drops safe for babies' eyes

Alexander Graham Bell (late 1800s) – invented the telephone

George Washington Carver (early 1900s) – discovered over 300 products that could be made from peanuts (ex: oil, cheese, soap, and coffee)

Nicolaus Copernicus (around 1500) – demonstrated that the Sun is the center of the solar system

Martha Coston (mid-late 1800s) – developed and manufactured maritime signal flares

Marie & Pierre Curie (late 1800s-early 1900s) – known for their work on radioactivity and discovery of radioactive elements

Charles Darwin (1800s) – traced the origin of man; explained theories of evolution in *The Origin of Species by Means of Natural Selection*

Thomas Alva Edison (late 1800s-early 1900s) – developed the electric light bulb, phonograph, mimeograph machine, and motion pictures

Albert Einstein (late 1800s) – discovered that mass can be changed into energy and that energy can be changed into matter; also known for his *Theory of Relativity*

Michael Faraday (early 1800s) – developed the first electric generator and electric motor

Alexander Fleming (early 1900s) – developed penicillin

Henry Ford (early 1900s) – built the first gasoline engine and the first assembly line to speed up the production of automobiles

Benjamin Franklin (mid 1700s) – proved that lightning is electricity

Sigmund Freud (late 1800s) – Austrian physician who established the field of psychoanalysis

Galileo Galilei (early 1600s) – formulated the *Law of Falling Bodies* and wrote about acceleration, motion, and gravity; developed the first astronomical telescope and made many discoveries in astronomy

William Harvey (1600s) – showed how blood circulates through the human body

Heinrich Hertz (late 1880s) – discovered electromagnetic radiation

Hippocrates (300-400 B.C.) – founded the first school of medicine; known as the *Father of Medicine*

Grace Hopper (mid-late 1900s) – developed new applications for computers; devised computer languages and language devices

Edward Jenner (late 1700s) – founded the science of immunology by developing a vaccine to protect the body against smallpox

James Joule (1840s) – showed that heat is a form of energy

Stephanie Kwolek (early–mid 1900s) – developed Kevlar, (a fiber stronger than steel) used in bullet resistant clothing

Carolus Linnaeus (mid-1700s) – a Swedish naturalist and physicist who devised a systematic method for classifying plants and animals

Barbara McClintock (mid-late 1900s) – Nobel Prize winner (1983) who did pioneer work with genes, studying cell structure in corn kernels

Gregor Johann Mendel (mid-1800s) – founded genetics through his work with recessive and dominant characteristics of plants

Navy Rear Admiral Grace Hopper, known for her advancements in computer technology, is often called the *Mother of the Computer*. She is probably the only woman scientist to have a ship named after her. The USS Grace Hopper DDG 70. She is also famous for the saying, *It's easier to ask forgiveness than it is to get permission.*

Dmitri Ivanovich Mendeleev (late 1800s) – developed the periodic table for classification of the elements in 1869

Sir Isaac Newton (1600s) – discovered that the force of gravity depends upon the amount of matter in bodies and the distances between the bodies; formulated the laws of gravity and motion

Louis Pasteur (mid 1800s) – developed a method for destroying disease-producing bacteria and for checking the activity of fermentative bacteria (pasteurization)

Elizabeth Pinckney (mid-late 1700s) – ground-breaking agriculturalist who had great success with growing indigo

Pythagoras (500s B.C.) – developed the Pythagorean theorem, which states that the sum of the squares of the legs of a right triangle is equal to the square of the hypotenuse

Jonas Salk (mid 1900s) – an American physician and bacteriologist who developed a vaccine to prevent polio

James D. Watson and Francis H. C. Crick (mid 1900s) – created a model of the molecular structure of DNA; won the Nobel Prize in 1962

James Watt (late 1700s) – invented the modern steam engine. The *watt*, a measure of electrical power, was named after him.

Eli Whitney (late 1700s) – invented the cotton gin and developed a faster way to make manufactured goods

Orville and Wilbur Wright (early 1900s) – succeeded at the first powered, controlled, and sustained airplane flight at Kitty Hawk, North Carolina on December 17, 1903

Looking for Answers

Scientists have a serious, consistent way of looking for answers about the world. They follow some steps that other scientists can repeat.

How Do Scientists Work?

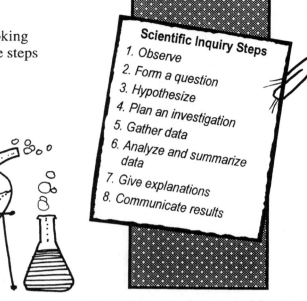

 They . . .

. . . observe carefully.

. . . take notes.

. . . ask questions.

. . . look at what already is known (do careful, thorough reading and research).

. . . examine and use knowledge from many branches of science.

. . . compare their questions and answers to knowledge already available.

. . . form hypotheses (careful, informed guesses) about what could be true.

. . . plan investigations that can test a hypothesis.

. . . follow clear procedures of scientific inquiry.

. . . control the variables that are not being examined in the investigation.

. . . use tools and instruments to collect and classify data.

. . . use tools and instruments to organize and summarize data.

. . . use math to measure, calculate, and express findings.

. . . keep careful records of their results.

. . . analyze and describe the results completely.

. . . use logic to understand relationships and draw conclusions.

. . . draw conclusions and develop explanations for the findings.

. . . provide evidence to support their conclusions.

. . . review and repeat the investigations.

. . . make their methods of study and results public.

. . . publish studies and results in scientific journals.

. . . review and repeat studies done by other scientists.

. . . keep records that are accurate and understandable.

. . . pay attention to new questions raised by the investigations.

Scientific Inquiry Steps
1. Observe
2. Form a question
3. Hypothesize
4. Plan an investigation
5. Gather data
6. Analyze and summarize data
7. Give explanations
8. Communicate results

MAC'S INQUIRY

1. OBSERVE

The water was too hot when Mac stepped into the bathtub, so he added some cold water. Right away the water cooled off. When it got too cool, he added hot water. The hot water did not seem to mix in very well. It took a lot of swishing and mixing before the tub warmed up.

2. QUESTION

Hmm, does hot water mix into cold water at a different rate than cold water mixes into hot water?

3. HYPOTHESIZE

Mac guessed that cold water mixes into hot water faster than hot water mixes into cold water.

4. INVESTIGATE

Mac filled ½ the bottle with cold water and added some food coloring. He filled ½ the other bottle with hot water and held cardboard over the opening. He turned the cold water bottle upside down over the other bottle, and pulled out the cardboard to let the waters mix.

I'll keep track of the time it takes for the blue color to spread throughout the bottle on the bottom!

5. GATHER DATA

Now I'll do this in reverse, adding colored hot water to the cold water.

I'll repeat the whole thing to see if I get the same results twice.

6. ANALYZE & SUMMARIZE

In both trials, it took about 3 seconds for the cold water to mix into the hot water. The hot water just stayed at the top of the bottle. It took over 15 minutes for the hot to mix into the cold.

7. EXPLAIN RESULTS
8. COMMUNICATE RESULTS

Cold water is denser than hot water, so it's harder for other molecules to mix in. Hot water molecules are less dense, so they stay on top of the cold water until I mix them in.

My chart shows the results.

Science Processes

As scientists work and wonder, they use certain processes (or skills of *doing*). You may not be a professional scientist, but if you are going to try to uncover facts, answer questions, or solve mysteries of the natural world, you'll need to be good at using these same processes.

Observing – using the senses to obtain information

If you pay attention to what your eyes, ears, nose, tongue, and skin are telling you, you might notice such things as . . .

. . . the heat in your palms when you rub your hands together quickly

. . . the strange smell that rises when you smash a cooked egg yolk

. . . the change in pitch of a siren sound as the speeding ambulance passes you

. . . the sediment that collects in the bottom of your glass of iced tea

. . . the difference in the taste of a green banana and a very ripe banana

Comparing – observing how things are alike or different

As you observe, keep a comparing attitude going at all times. Constantly match up one event against another. For instance, you might notice . . .

. . . that all the body joints are located at places where bones come together.

. . . that all body joints allow movement of some kind.

. . . that some joints rotate, while others bend like a hinge, swivel, or twist.

Questioning – raising uncertainty or making a statement that seeks an answer

Observations often raise questions. For instance . . .

. . . Why does a cooked egg spin smoothly and a raw egg wobble all over?

. . . Will a magnet attract all objects made of metal?

. . . Does the color of the glass affect the rate at which the ice melts?

Hypothesizing – tentatively accepting an explanation as the basis for further investigation

Observations can lead to a smart guess about what might happen in an investigation. A hypothesis should be made carefully. It must be a statement that can be tested in an investigation or experiment.

Can be tested:

. . . The speed of a ball rolling down an incline will increase as the incline angle increases.

. . . Hot water freezes faster than cold water.

Can't be tested:

. . . Salt and water form a solution faster than any other two substances.

64

Better Grades & Higher Test Scores / SCIENCE
Copyright ©2003 by Incentive Publications, Inc., Nashville, TN.

Experimenting – testing under controlled conditions

Controlling Variables – managing factors that may influence an experiment

An experiment should be carefully planned and controlled. It is important to manage any factors which might influence the outcome. Only the factor being tested should be allowed to vary. The other factors (variables) should be controlled to remain the same. For instance . . .

Suzy does an experiment to find out which kind of liquid popsicle will freeze the most quickly. The variable is the five different liquids (water, cream, root beer, chocolate milk, and soy milk). It is important that the exact same amount of each liquid be tested. Also, all liquids must be placed in containers of the same size and same material. The containers should be placed in the same freezer in the same location. The same instruments for measuring time and temperature should be used for all samples.

All the variables are controlled except the kind of liquid — same amount, same containers, same freezer, and they go in at the same time.

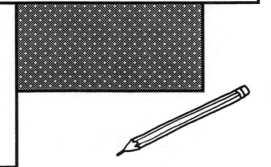

Classifying – assigning objects or processes to a category based on a common characteristic

Events, objects, or processes belong to one or more groups or classes. When you group something, make sure it shares the characteristics of the other items in the group. Here are some examples . . .

. . . Vertebrates can be divided into classes: fish, reptiles, mammals, amphibians, and birds.

. . . Copper, neon, wood, water, nitrogen, iron, and sand are all substances. But copper, neon, nitrogen and iron can be classified as elements. The others are not elements. Wood, water, and sand are compounds.

. . . These characteristics all fit into a group called life processes: osmosis, respiration, reproduction, metabolism, mitosis, photosynthesis, diffusion. All these processes can also be called plant processes, because they all take place in plants. Photosynthesis does not take place in animals, so these cannot all be classified as animal processes.

Defining Operationally – listing the characteristics or behaviors by which something is defined

Here is an operational definition of a liquid . . .

A liquid is a substance that can be poured. It has no shape of its own, but takes on the shape of whatever container into which it is placed.

Get Sharp: Science Processes

Formulating Models – devising a concrete representation to illustrate an abstract idea or relationship, or to show something that cannot be seen easily

A model can be a physical (3-dimensial) creation, a picture or diagram, or a mathematical formula. Models can be built, drawn, designed on a computer, or written. For example . . .

. . . Spherical objects and a source of light can be used to model the solar system.

. . . CO_2 is a mathematical model of the compound carbon dioxide.

Using Math – using numbers to count, measure, or otherwise give quantities of data

Science and math are close friends. Math is needed for just about every science question or investigation. For instance, you'll use math to . . .

. . . count the number of fruit flies hatched since yesterday.

. . . measure the distance you tossed each of those eggs.

. . . find the difference in the height of two plants you're growing.

. . . comparing the temperatures at which two substances melted.

. . . calculate the speed of a pendulum swing.

Summarizing & Recording Data – systematically expressing the results of collected data and storing it in an organized fashion

The data found in an investigation need to be recorded. The results need to be organized in a way that they make sense as a part of the record. The record might be written, typed into a computer, recorded on an audio or videotape, or represented some other way.

Why do I have to keep a record? Isn't *doing* it enough?

Because . . .

Recording helps you see the results.

A good record makes it possible for others to follow your steps and repeat your investigation!

Others can learn from your record.

This helps you remember what you did, and helps you explain what you found.

Recording shows others what you did.

Interpreting Data –
finding patterns or relationships in a set of data

When you look at information gathered from an investigation, search for patterns in the data and look for relationships among the events. For example . . .

I notice that all the substances that were attracted to these magnets were made of metal.

The thicker liquids took a longer time to freeze.

Inferring – implying a conclusion from available evidence

So what does the evidence (data) tell you? When your investigation is finished, you'll need to come to some conclusion about what you've found.

Since I tasted sweet, sour, bitter, and salty on different areas of my tongue, I infer that all taste buds are not sensitive to the same tastes.

Predicting – foretelling from previous information

Sometimes the results of an investigation give you data that make it possible for you to predict events or results you haven't seen.

When I left celery in a glass of red punch, the celery turned red. I predict this daisy will turn blue if I leave it in a glass of blue water.

Communicating – exchanging information

New information or results from an investigation don't do much good if they're not shared. Practice ways of communicating what you've learned. These are just a few ways . . .

drawing	model	graph	written log
videotape	article	poster	book
typed record	tape recording	email	table or chart

Get Sharp: Science Processes

Safe Practices in the Science Lab

It can be fun, amazing, or fascinating to investigate science questions. It can also be hazardous. Practicing science safely is a serious matter. The equipment, substances, and processes all need to be handled with great care. If you follow accepted practices for science safety, you'll discover more, and have plenty of excitement without danger.

1. Always get permission from a teacher or adult before starting an investigation on your own.
2. Review the lab procedure before you start, so you know what you'll be doing.
3. Keep a fire blanket, fire extinguisher, first aid kit, and eyewash kit nearby.
4. Use the safety equipment. Wear goggles and an apron during experiments.
5. Only use lab equipment. Don't do experiments with utensils used elsewhere.
6. Tie back loose hair and clothing when working in the lab.
7. Be careful to keep excess material away from flames.
8. Always slant a test tube away from you when you are heating a substance.
9. Never inhale materials or chemicals.
10. Never taste substances or chemicals.
11. Never eat out of containers in the lab. (In fact, don't eat in the lab at all!)
12. Treat glass objects very carefully.
13. Handle all equipment as if it were hot or dangerous.
14. Be extra cautious around chemicals, sources of heat, open flames, fumes, or electrical equipment.
15. If you spill anything on your clothes or skin, wash it off with lots of water—immediately.
16. If anything gets in your eye, wash it with lots of water—immediately.
17. If your clothes catch on fire, DO NOT RUN. Smother a fire under a coat or fire blanket.
18. Flush any burns with cold water—immediately.
19. Tell the teacher about all spills, splashes, accidents or injuries—right away.
20. Return all substances to their original containers. Close containers tightly.
21. Do not wash any of the substances down the sink. Dispose of them as you are directed.
22. Before you leave the lab, make sure all sources of heat are turned off and all electrical devices are disconnected.

TOP 10 BAD IDEAS

FOR SCIENCE LAB BEHAVIOR

Number 10: Taste the chemicals to see if you can identify them.

Number 9: Insist to yourself (and everyone else) that you can see better without your safety goggles.

Number 8: Don't worry if your charming new hairdo gets in the way of your experiment.

Number 7: Wipe up all chemical spills with your shirt sleeve.

Number 6: Take good strong whiffs of the chemicals you are heating or mixing.

Number 5: Always slant the test tube toward you when you are heating or mixing substances.

Number 4: Use an empty test tube as a goblet for drinking lemonade.

Number 3: Get a brilliant notion to use your Bunsen burner for popping corn.

Number 2: Pour leftover chemicals down the sink.

The **Number 1 Bad Idea:** If your clothes catch on fire - *RUN!*

69

The Big Ideas of Science

If you study science, you will keep running into these big ideas. No matter what specific topic, idea, or field you dabble in, these will pop up. That's because there are some basic underlying concepts that relate to or explain the natural world.

Everywhere I look, I see another system!

Systems

A system is an organized group of related components or objects that form a whole and work together as a whole to perform one or more functions. Each system has components and boundaries. A change in one component of a system affects the whole system. Most systems function with an input and an output of materials and energy. Most systems are related or connected to other systems.

Examples:
an ecosystem	any organism	a cyclone
the solar system	a pulley	a river system
a digestive system	an engine	an electrical circuit

Organization

Organization is an arrangement of independent items, objects, organisms, units of matter, or systems joined into one whole system or structure.

Examples:
Living things are organized into several kingdoms.
The planets are organized in a particular order from the sun.
Elements are organized by the structure of their atoms.

Order

Order is the predictable behavior of objects, units of matter, events, organisms, or systems.

Examples:
When an animal cell is fertilized, it can be expected to start dividing.
Year after year on Earth, fall follows summer and spring follows winter.
As a substance gets cooler, the molecules move more slowly and they stay closer together.

Cycle

A cycle is a series of events or operations that regularly occur and usually lead back to the starting point.

Examples: life cycle of every living organism
rock cycle	oxygen-carbon dioxide cycle
life cycle of stars	movement of planets
water cycle	rising and falling and changing of tides
nitrogen cycle	phases of the Moon

70

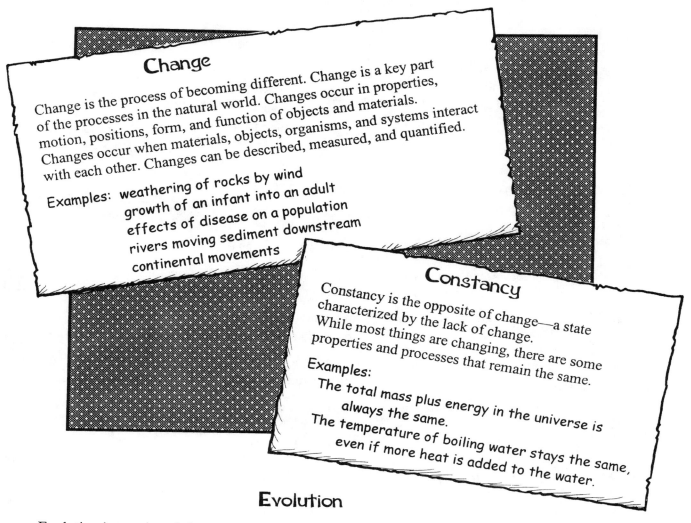

Change

Change is the process of becoming different. Change is a key part of the processes in the natural world. Changes occur in properties, motion, positions, form, and function of objects and materials. Changes occur when materials, objects, organisms, and systems interact with each other. Changes can be described, measured, and quantified.

Examples: weathering of rocks by wind
growth of an infant into an adult
effects of disease on a population
rivers moving sediment downstream
continental movements

Constancy

Constancy is the opposite of change—a state characterized by the lack of change. While most things are changing, there are some properties and processes that remain the same.

Examples:
The total mass plus energy in the universe is always the same.
The temperature of boiling water stays the same, even if more heat is added to the water.

Evolution

Evolution is a series of changes that cause the form or function of an object, organism, or system to be what it currently is. Things in the natural and man-made world change over time. The present forms and functions of systems, organisms, objects, and materials result from the past.

Examples: The universe as a system is constantly changing with the birth and death of stars.
After repeated exposure to certain antibiotics, some bacteria evolve to become resistant to the antibiotic so they are no longer affected by it.

Equilibrium

Equilibrium is the state in which equal forces occur in opposite directions and offset or balance each other. Most systems and interacting units of matter tend toward a state of balance— a steady state in which energy is uniformly distributed.

Examples: When the body temperature gets too high, the body sweats to help regulate the temperature back to normal so the body doesn't overheat.
When there is too much pressure in an area beneath Earth's surface, a volcano may erupt to release and equalize the pressure.

71

Form & Function

The shape or structure (form) of an organism, object, or system is often related to the operation (function). Frequently the function of something is very dependent upon its form.

Examples: Form: The wall of a plant cell is very sturdy.
  Function: This allows the plant cells to form sturdy stems or trunks.
  Form: The snail has a sticky, muscular foot.
  Function: This foot helps the snail to hold on to surfaces and pull its way along.
  Form: A bobsled has a very slim, sleek, bullet-like shape and a smooth surface.
  Function: This allows the bobsled to travel very fast with little wind resistance.

Cause & Effect

A cause is anything that brings about a result. An effect is the result (the event or situation or behavior) that follows from the cause.

Examples: Cause: An infection breaks out in the body.
  Effect: The body's defenses start to fight the infection.

  Cause: A predator comes close to a rattlesnake.
  Effect: The snake signals a warning by shaking its tail.

Energy & Matter

Energy and matter are closely related. They are constantly interacting with each other. This means that they affect one another or work together. Energy can move or change matter. Matter can be changed into energy. Energy can be transferred to matter.

Examples: The heat of sunlight causes ice cubes to melt.
  When wood burns, the wood is changed and heat is released.
  The energy of moving ocean water moves the sand along the shore.

Force & Motion

Force and motion have a close relationship. An object changes position, direction, or speed when a force acts on it.

Examples: Muscle power pushing down on a screwdriver
  lifts a tight lid up from a can.
  A woman in a canoe pulls her paddle backwards in the water.
  The canoe moves forward.
  A pitcher pitches a baseball to the batter.
  The batter swings. The ball collides with the bat,
  and changes direction, flying over the infield.

GET SHARP →

on Space Science

The Solar System

What a strange and wonderful community—one star, nine planets, many moons, thousands of asteroids and comets, and billions of other bits of rock! This is our solar system—the place we call home. A *solar system* is a star and a group of planets that revolve around it. Ours is not the only solar system in the universe. There are millions more.

The centerpiece of this solar system is our Sun, a brilliant star. Nine planets revolve around the Sun. Many of these planets have natural satellites that revolve around them. Between the paths of Mars and Jupiter, thousands of asteroids orbit the Sun in an asteroid belt.

Comparative Sizes of the Planets (Distances are not proportional.)

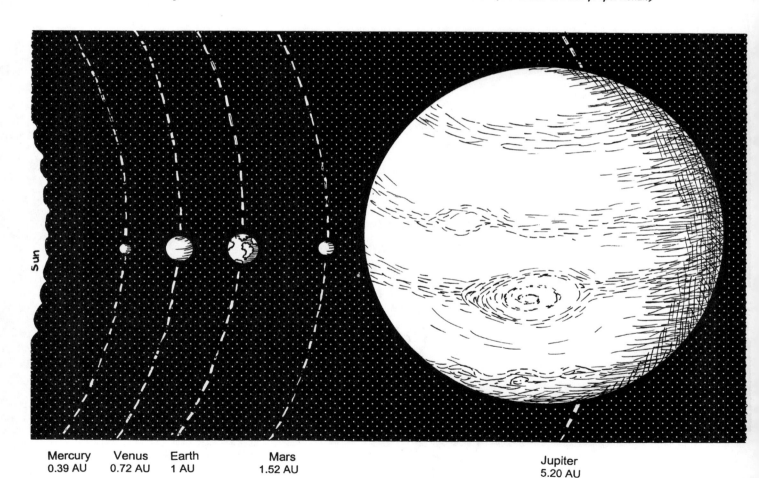

Mercury	Venus	Earth	Mars	Jupiter
0.39 AU	0.72 AU	1 AU	1.52 AU	5.20 AU

Distance from the Sun in Astronomical Units

Get Sharp: The Solar System

Better Grades & Higher Test Scores / SCIENCE
Copyright ©2003 by Incentive Publications, Inc., Nashville, TN.

The Sun's huge size gives it great gravitational pull. This is the force that holds the solar system together and keeps the planets moving on their paths around the Sun.

Each planet revolves around the Sun in a path called an *orbit*. In this solar system, the orbits are elliptical in shape. All the orbits are just about in the same plane, except for Pluto's, which varies by about 17 degrees.

Since the orbits are elliptical, each planet is not always the same distance from the Sun as it travels. The *perihelion* is the point in the orbit where the body is closest to the Sun. The *aphelion* is the point where the object is farthest from the Sun.

In 2001, astronomers discovered a large celestial body in the solar system beyond Pluto. They named it *Quaoar*. It is referred to as the largest and most distant minor planet of the solar system. In fact, with a diameter of 1250 km, *Quaoar* is the largest solar system object discovered since the 1930 discovery of Pluto.

Saturn	Uranus	Neptune	Pluto
9.54 AU	19.19 AU	30.1 AU	39.5 AU

75

The Sun

It's just one of millions of stars in the Milky Way galaxy. Although it is a huge, brilliant star, it is not unusual. Nevertheless, it is the most important star in our lives. We call it our Sun—this hot ball of gas that lies at the center of our solar system.

The Sun is a restless flame that is always changing. It bubbles and spurts, and hurls hot gases thousands of miles beyond its surface. It is a great nuclear furnace. In its core, nuclear fusion changes hydrogen to helium. These reactions produce energy, which radiates outward from the core. As the energy travels outward from the center, it is absorbed in the layers. By the time the energy reaches the surface, the heat has been reduced and temperatures are considerably lower.

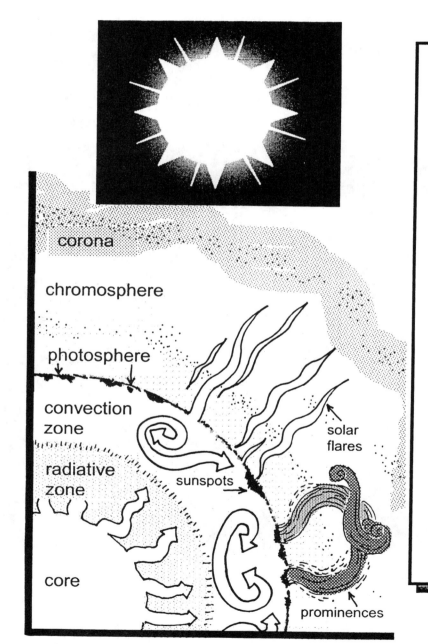

corona

chromosphere

photosphere

convection zone

radiative zone

sunspots

solar flares

core

prominences

Distances are not proportional.

Sun's Vital Statistics

Temperature at the center:
about 30 million° F

Temperature at the surface:
about 10,000° F

Diameter:
almost 900,000 miles

Age:
thought by some to be about 4.5 billion years old

Density at center:
over 150 X density of water

Distance from Earth:
93 million miles

Present composition:
70% hydrogen, 28% helium, 2% other elements

Also of interest:
- The Sun's light takes 8 minutes to reach Earth.
- The composition of the Sun changes over time as hydrogen is converted to helium in the core of the Sun.
- The Sun converts four million tons of gases into energy every second.

Occasionally, solar flares excite gases in the upper atmosphere of Earth, causing them to radiate some spectacular displays of colored lights in the night sky. These lights are called the *aurora borealis* in the Northern Hemisphere and the *aurora australias* in the Southern Hemisphere.

Cool dudes won't look directly at the sun, even wearing eye protection.

The Sun's Composition & Structure

Core – Heat and light are produced in this center of the Sun by nuclear fusion reactions.

Radiative Zone – Energy from the reactions flows out in waves from this area beyond the core to the convection zone.

Convection Zone – The waves reach a place where they no longer have the strength to keep pushing forward. In this region, the waves churn around, and the heat travels on by convection.

Photosphere – This is the name given to the visible surface of the Sun. It emits the radiation that we can see from Earth.

Chromosphere – This bright red layer of gas extends about 6000 kilometers above the photosphere. The chromosphere can only be seen during an eclipse.

Prominences – These "fingers" of flame are surges of glowing gas that rise from the Sun's surface and shoot outward from the chromosphere.

Corona – This outer atmosphere of the Sun is a transparent region beyond the chromosphere that extends millions of miles into space. It is visible from Earth only during eclipses.

Solar Flares – These are sudden increases in brightness of the chromosphere, occurring near sunspots. Electrons and protons leap out from these areas at high speeds, sometimes reaching and affecting Earth's atmosphere.

Sunspots – These are relatively cooler areas of the Sun that look dark by comparison. Sunspots develop and may last for a day, or for as long as several months. They can measure thousands of miles in diameter.

Solar Wind – The Sun gives off a low-density stream of charged particles known as the solar wind. The wind can affect the borealis, the tails of comets, and the travels of spacecraft.

Solar energy is a renewable resource. Solar panels can collect the sun's rays, and solar cells can convert the light into electricity.

Get Sharp: The Solar System

Planetary Particulars

A *planet* is a large body that orbits a star.
In our solar system, we know of nine planets that orbit our sun.

The four inner planets (Mercury, Venus, Earth, Mars) are referred to as the **terrestrial** (or rocky) planets. They are composed mostly of rock and metal. In general, these planets have slow rotation, solid surfaces, high densities, few satellites (or moons) and no rings.

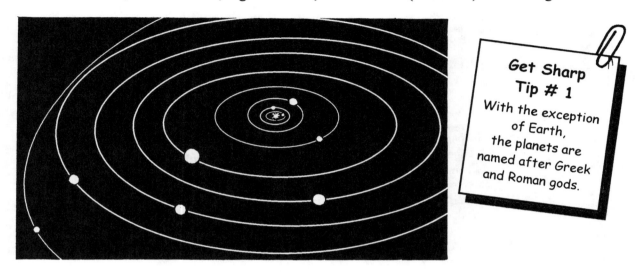

Get Sharp Tip # 1
With the exception of Earth, the planets are named after Greek and Roman gods.

The Inner Planets

	Mercury	Venus	Earth	Mars
Average distance from the Sun (in miles)	36 million	67 million	93 million	142 million
Diameter at the equator (in miles)	3049	7565	7926	4220
Mass (times Earth)	.06	.82	1 (6.6 sextillion tons)	.11
Length of Day/Night (Rotation time in Earth time)	176 days	117 days	23 hours, 56 minutes	24 hours, 37 minutes
Revolution time (In Earth time)	88 days	225 days	365 days, 5 hours	687 days
Surface Gravity (Earth = 1.0)	0.38	0.9	1.0	0.38
Average Temperature or Temperature Range	-300° to -800°F	900°F	-136° to 128°F	-116° to 32°F
Number of Satellites or Moons	0	0	1	2

Better Grades & Higher Test Scores / SCIENCE

Four of the five outer planets (Jupiter, Saturn, Uranus, and Neptune) are referred to as the gas planets. They are composed mostly of gases and some liquid. In general, these planets have rapid rotations, low densities, rings, and many satellites (or moons). Pluto's composition, which is mostly dust and ice, is unlike any of the other planets.

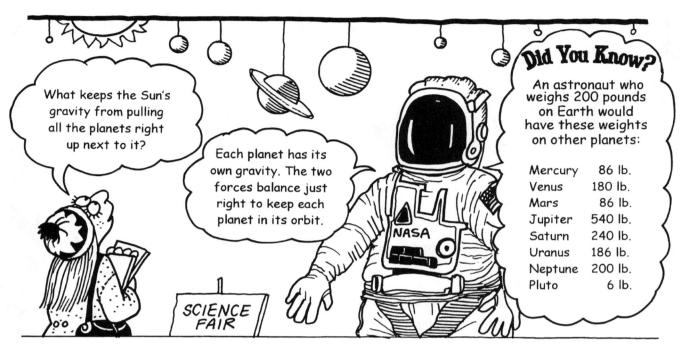

The Outer Planets

	Jupiter	Saturn	Uranus	Neptune	Pluto
Average distance from the Sun (in miles)	484 million	887 million	1.8 billion	2.8 billion	3.7 billion
Diameter at the equator (in miles)	89,500	75,000	31,135	30,938	1,875
Mass	317.9	95.2	14.54	17.2	.002
Length of Day/Night (Rotation time in Earth time)	9 hours, 50 minutes	10 hours, 14 minutes	17 hours, 14 minutes	17 hours, 6 minutes	6 days, 9 hours
Revolution time (In Earth time)	11.9 years	29.5 years	84 years	165 years	248 years
Surface Gravity (Earth = 1.0)	2.7	1.2	0.93	1.5	0.03
Average Temperature or Temperature Range	-238° F	-274°F	-328°F	-346°F	-380°F
Number of Satellites or Moons	16	18	15	8	1

Better Grades & Higher Test Scores / SCIENCE
Copyright ©2003 by Incentive Publications, Inc., Nashville, TN.

Get Sharp: The Solar System

The Inner Planets

MERCURY - The Speediest Planet

Mercury travels at 29.7 miles per second in its orbit.

Vital Statistics

• the first planet, closest to the Sun
• dark-colored, rocky planet
• second smallest planet
• one of the most elliptical orbits of all planets
• fastest planet in the solar system
• shortest year (revolution time)
• longest day (rotation time)
• no moons or other satellites
• extremely thin atmosphere
• strong magnetic field
• heavily cratered surface and wrinkled crust
• extreme hot and cold temperatures
• second densest planet

Also of interest:

• A large crater called the *Caloris Basin* is one of the largest surface features on Mercury. It was probably created by an impact with some other body in the solar system.

Mercury is the god of commerce, travel, and thieves. Mercury is also the messenger of the Roman gods.

See other vital statistics on page 78.

VENUS - The Visible Planet

Venus is usually visible from Earth without a telescope.

Venus is the Roman goddess of love and beauty.

Vital Statistics

• sixth largest planet
• second planet from the Sun
• nearly circular orbit
• dense carbon dioxide atmosphere
• hotter than Mercury
• slow, retrograde rotation
• mostly rock, with surface of mountains, volcanoes, large craters, and lava plains
• extremely dense atmosphere
• no moons or other satellites
• atmosphere traps Sun's radiant heat

Also of interest:

• Sulfuric acid drips from the clouds giving the planet an orange color.
• Venus may have had water, but it all boiled away from the heat.

See other vital statistics on page 78.

EARTH - The Home Planet

Earth is only planet known to have an atmosphere that can sustain life.

Vital Statistics

• third planet from the Sun
• fifth largest planet
• densest major body in the solar system
• solid inner iron core and rocky crust; crust varies in thickness
• semi-fluid outer core and mantle layers
• plants release oxygen into the atmosphere
• atmosphere is 78% nitrogen, 21% oxygen, 1% other gases
• ozone layer absorbs dangerous ultraviolet rays from the sun
• only planet with liquid water on surface
• one natural satellite

Also of interest:

• Humans have placed thousands of artificial satellites into orbit around Earth.
• Earth could not be fully explored until spacecraft could be launched to see it from space.

See other vital statistics on page 78.

Earth is the only planet whose name does not come from Greek or Roman mythology. It is of Germanic and Old English origin.

MARS - The Red Planet

A rusty surface coating is blown around by great storms, making the planet look red.

Mars is the Roman war god.

Vital Statistics

• fourth planet from the Sun
• seventh largest planet
• highest mountains and deepest valleys in solar system
• permanent ice caps at the poles of frozen carbon dioxide
• winds raise huge dust storms at the surface
• thin atmosphere, mostly of carbon dioxide
• covered with a thick cloud layer
• surface of craters, mountains, valleys, volcanoes
• two natural satellites
• highly interesting and varied terrain

Also of interest:

• Many scientists believe there may be traces of liquid water on Mars.
• Mars is a bright planet, visible on Earth when it is in the nighttime sky.

See other vital statistics on page 78.

The Outer Planets

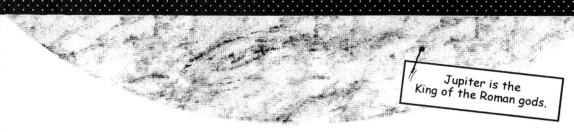

JUPITER – The Giant Planet

Jupiter is the King of the Roman gods.

All the other planets could be squeezed into Jupiter.

Vital Statistics

- largest planet, fifth planet from the Sun
- fastest-spinning planet (day lasts less than 10 hours)
- fourth brightest object in the sky
- a gas planet—90% hydrogen, 10% helium
- three layers of icy clouds
- radiates heat from an internal heat source
- strong magnetic field
- has a great red spot *(probably caused by a violent storm)*
- dark rings, probably of rocky particles
- 16 satellites that have been named
- largest moon in the solar system *(Ganymede)*

Also of interest:

- The fast rotation pulls the planet slightly out of shape with a bulge at the equator.
- High winds blowing in opposite directions cause light and dark colored bands that are features of Jupiter's vividly colored appearance.

See other vital statistics on page 79.

SATURN – The Ringed Planet

Saturn has the largest and most complicated system of rings.

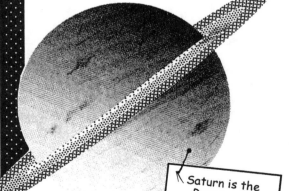

Saturn is the Roman god of agriculture.

Vital Statistics

- second largest planet
- over 1000 rings formed from icy particles
- sixth planet from the Sun
- flattened from fast rotation
- least dense planet
- composed of 75% hydrogen, 25% helium
- strong magnetic field
- radiates heat from an internal heat source
- 18 named satellites
- a great brown spot near the north pole

Also of interest:

- Saturn could float on water. It is less dense than water.
- Every 30 years or so, violent storms cause bright white spots on the surface.
- The origin of Saturn's rings is unknown.

See other vital statistics on page 79.

URANUS - The Tilted Planet

The axis of Uranus is tilted so far that the planet virtually orbits on its side.

Vital Statistics

- seventh planet from the Sun
- third largest planet
- discovered by accident in 1871
- retrograde rotation
- 11 known dark rings of rock, dust, and ice
- 15 known satellites
- moons are named for Shakespearean characters

Also of interest:

- Because of the tilt, each pole stays in darkness for 40 years.
- The blue color is caused when methane in the atmosphere absorbs red light.

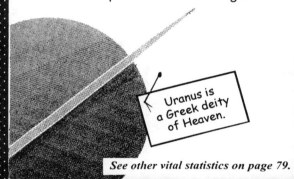

Uranus is a Greek deity of Heaven.

See other vital statistics on page 79.

NEPTUNE - The Windy Planet

Neptune has the fiercest winds in the solar system, blowing up to 1400 miles per hour.

Neptune is the god of the sea in Roman mythology.

Vital Statistics

- eighth planet from the Sun
- fourth largest planet
- gaseous planet
- radiates heat from an internal heat source
- brilliant blue methane atmosphere
- three thin, dark rings around the planet
- 8 known moons

Also of interest:

- The Great Dark Spot changes size and shape every 16 days.
- Every 250 years, Neptune is the outermost planet when Pluto comes closer to the Sun for 20 years at a time.

See other vital statistics on page 79.

PLUTO - The Tiny Planet

Pluto is smaller than Earth's moon, and by far the tiniest planet in the solar system.

Vital Statistics

- farthest planet from the Sun
- most recently-discovered planet
- the most elliptical orbit
- seemingly a snowball of frozen gases
- thin atmosphere
- severely tilted axis (over 62°)

Pluto is the Greek god of the underworld.

Also of interest:

- Pluto is the only planet never visited by a spacecraft. It is too far away even to be seen clearly by telescopes.
- Pluto is so similar in size to its moon, Charon, that many astronomers consider them a double planet.

See other vital statistics on page 79.

83

Other Solar System Objects

The solar system contains thousands of other interesting objects in addition to the Sun and the nine planets. Most of them are smaller than the planets. These objects are classified as *asteroids* or *comets*, and sometimes are called *minor planets* or *planetoids*. Although much of the space between planets seems empty, there are billions of particles of dust (much of it microscopic) floating around. This is often referred to as *space dust* or *cosmic dust*.

> The asteroid with the orbit closest to Earth is Hermes. It comes within about 500,000 miles of Earth.

> The largest asteroid, Ceres, has a diameter of 620 miles.

Asteroids

Asteroids are small, dense, rocky objects that orbit the sun. They are fragments of material similar to the inner planets, ranging in size from very tiny to miles across. Most asteroids orbit in a belt of 100,000 or more between Mars and Jupiter. The largest asteroids are named, for example: *Ceres, Psyche, Juno*, and *Davida*.

Meteoroids

Very small asteroids orbiting the sun are called *meteoroids* to distinguish them from larger asteroids. Some meteoroids are particles thrown off from the nuclei of comets.

Meteors, Shooting Stars, & Fireballs

Millions of meteoroids enter Earth's atmosphere every day. When this happens, the air friction heats the meteoroid, creating a glowing trail of gases. When a streak of a meteoroid is visible in the sky, it is called a *meteor*. Because the streak of flashing light is quite spectacular, meteors are often called *shooting stars* or *falling stars*. An unusually bright meteor is called a *fireball*.

Meteor Showers

When Earth passes through a large group of meteoroids, several may hit Earth's atmosphere. This can cause a number of meteors to be seen at once, resulting in a *meteor shower*.

Meteorites

Generally, meteors burn up in Earth's atmosphere. When a piece of a meteor survives the trip through the atmosphere and collides with Earth's surface, it is called a *meteorite*.

> The largest meteorite ever found fell in Namibia. It weighs about 60 metric tons.

Comets

A *comet* is a large clump of ice, dust, and frozen gases that travels around the Sun in a long orbit. As the body nears the Sun, some of the ice melts or vaporizes. Streams of gases and particles fly away from the main body of the comet and create a spectacular tail that shines in the sunlight and stretches for miles.

The nucleus is the mostly solid part of the comet. This consists of ice and gas with solid particles and dust.

The coma is a thick cloud of water, carbon dioxide, and other gases surrounding the nucleus. Though a nucleus may be only a few miles across, a coma can be wider than a planet.

The tail is the most spectacular part of a comet. It is made of dust and gas, and can trail along behind the comet for millions of miles. When the sunlight shines through the trail, it creates a dazzling sight!

The orbit of a comet is an eccentric path that goes far beyond the farthest planet and back around the Sun. The orbit of a comet is usually not in the same plane as the planets and Sun, but is at an angle to the orbits of the planets. Orbits vary greatly in length. Some comets take thousands of years to complete an orbit. Comet Encke has a shorter orbit, taking only about three years to complete the path.

The most famous comet is Comet Halley. It reappears every 76 years. Its last appearance was in 1985.

The Weekly Planet   1997

Comet Dazzles Skywatchers

This week, sky watchers got a rare treat. Comet Hale-Bopp paid a visit, coming within 122 million miles of our planet. With its long, spectacular tail throwing off large amounts of dust and gas, the comet was easily visible from Earth.

A comet can age! Each time a comet orbits the Sun, it loses some of its ice and dust. Eventually, a comet may lose its entire nucleus. It may break up into a bunch of asteroids, or turn into a group of meteors.

A comet is named after the person who discovers it.

Comets are often called dirty snowballs or icy mudballs.

Get Sharp Tip # 2

Comets orbit the Sun. They are only visible when they get near the Sun.

Earth's Moon

The Moon is the brightest and biggest thing in the night sky. Sometimes it is a huge, glowing ball. Sometimes it seems to disappear. It is beautiful, romantic, shadowy, eerie, and mysterious. Its changes, the marks on its surface, and its shiny beauty have fascinated people for centuries. What is the Moon, our closest neighbor?

The Moon is the one natural satellite of our planet. It is a rocky, cratered body orbiting Earth and shining from the reflected light of the Sun. Its topography is varied—with hills, mountain ranges, valleys, chains of craters, ridges, faults, and waterless "seas."

The Moon's Vital Statistics

Diameter: 2160 miles
Distance from Earth: 221,000 - 252,000 miles
Orbit time: 27.3 days
Temperature: -274° F to 230° F
Day/Night: 29.5 days

Also of interest:

• surface covered with fine dust and rocky bits
• no wind
• no weather
• no atmosphere
• extreme variation in temperatures
• craters formed by collisions with other bodies

The gravitational forces between Earth and the Moon cause some interesting effects. One of these is tides in Earth's waters *(See page 122.)* Gravity also controls the rate of the Moon's rotation, and is responsible for the fact that only one side of the Moon is ever seen from Earth.

The Moon rotates once during its revolution. Because of this, the same side always faces Earth. The back side of the Moon was never seen until a Soviet spacecraft orbited the Moon in 1959.

The date was July 20, 1969, when the U.S. spacecraft *Apollo 11* set down on the Moon's Sea of Tranquility. Two American astronauts walked on the Moon's surface that day.

Since the Moon has no wind or weather, their footprints may never be disturbed.

Why does the Moon seem to have a face?

The "features" are actually part of the Moon's surface. Some of these are rocky craters, volcanoes, and shadowy pit areas known as seas. (They are seas that have never had any water!)

The Moon is the only extraterrestrial body visited by humans.

As the Moon orbits Earth, the positions of the Earth, Sun, and Moon change in relationship to each other, so the Moon looks different on different nights. These changes that we see are known as the **phases of the Moon**. The Moon takes about 29 ½ days to complete its revolution cycle.

The Moon's Phases

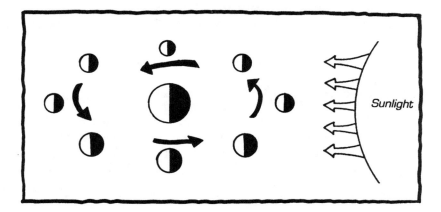

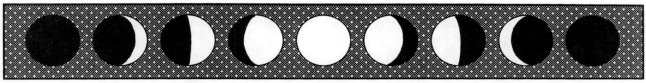

| new moon | waxing crescent | 1st quarter | waxing gibbous | full moon | waning gibbous | 3rd quarter | waning crescent | new moon |

Lunar Eclipse

An *eclipse* occurs when the shadow of one space body falls on another body, or when one blocks the light shining on another body. In any eclipse, one space body is darkened. A *lunar eclipse* occurs when the Moon is darkened because it is passing through Earth's shadow. Earth's shadow is made up of two parts: a dark *umbra* and a fainter *penumbra*. A total eclipse can only be seen when the Moon is covered by the umbra. Even during a total lunar eclipse, the Moon does not disappear completely from sight. Usually it appears to be dark red, because some of the Sun's light passes around the edges of Earth and reflects off the Moon.

A lunar eclipse does not happen every month. As the Moon orbits Earth, it often passes below or above Earth's shadow. An eclipse only occurs when the Moon, Earth, and Sun are positioned in a straight line during a full Moon phase. A lunar eclipse can be seen from the night side of Earth. It can last up to an hour and forty-five minutes.

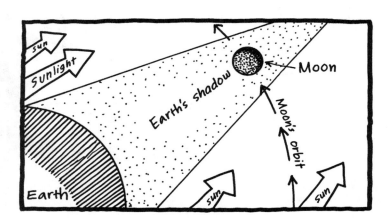

> **Get Sharp Tip # 3**
> A **total eclipse** occurs when the entire Moon passes through Earth's shadow. If only a part of the Moon passes through the shadow, the eclipse is **partial**.

87

Solar Eclipses

It's the middle of the day. The sun is disappearing, and the sky is darkening. Imagine how frightening this must have been for ancient people who did not understand the movements of celestial bodies! Now we know that this strange occurrence is a solar eclipse, and we understand why it happens. A *solar eclipse* occurs when the Moon's shadow falls on the Earth.

There are three kinds of solar eclipses.

Total Eclipse

A *total eclipse* is the most unusual and spectacular kind of solar eclipse. People will travel long distances to catch a glimpse of one, because this is the only time the Sun's corona can be seen without special equipment. And the Sun's corona is quite a sight! As the Moon's shadow moves from west to east across the Earth, the Moon completely blocks the Sun. At that moment, the Sun's outer atmosphere, the corona, is visible for a short time. It flashes around the darkened circle of the Sun. Sun flares may be seen also during a total eclipse.

A total eclipse is brief. The average length is about two and one-half minutes. Some have lasted as long as seven minutes, but many last only a few seconds. A total eclipse can only be seen in the area of the Earth that is covered by the umbra, or the inner shadow cast by the Moon.

Partial Eclipse

A *partial eclipse* occurs when the Moon only covers part of the Sun. A partial eclipse is seen in areas of Earth where only the penumbra, or fainter shadow, is cast on Earth.

Annular Eclipse

An *annular eclipse* may happen when a total eclipse occurs while the Moon is at its farthest point from the Earth in its orbit. In this situation, the Moon does not quite cover the Sun, and the shadow only covers the center of the Sun. A ring of the Sun is visible around the darkened center. The corona cannot be seen during an annular eclipse.

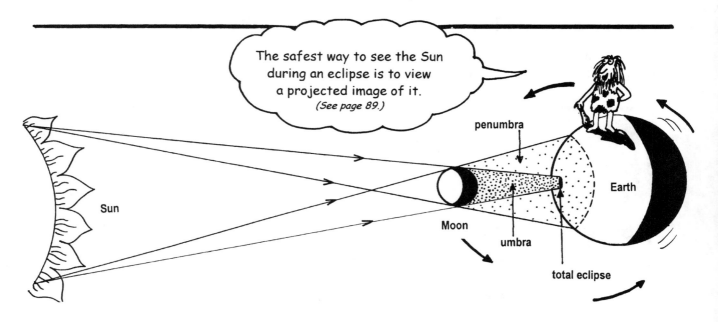

The safest way to see the Sun during an eclipse is to view a projected image of it.
(See page 89.)

penumbra

Sun

Moon

umbra

Earth

total eclipse

Better Grades & Higher Test Scores / SCIENCE
Copyright ©2003 by Incentive Publications, Inc., Nashville, TN.

NEVER LOOK THE SUN IN THE FACE (or How to View the Sun Safely)

1. First, choose a box about two feet long and cut a one-inch hole in the end.

2. Tape a piece of white paper to the INSIDE of the end opposite the hole.

3. Tape an index card over the OUTSIDE of the hole. Use a pin to punch a hole in the center.

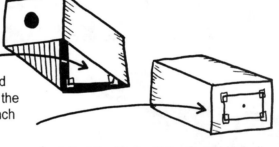

4. Adjust the box until the Sun shines directly through the pinhole and projects a glowing disk on the far end. Keep watching until the Moon passes across the Sun, causing the solar eclipse.

Get Sharp Tip # 4
It may take a long time for the Moon to completely block the Sun. Protect yourself from the Sun's harmful rays with sunblock lotion and appropriate clothing and hats.

89

The Milky Way & Beyond

Our Sun is just one of billions of bright stars in space. With its planets, our Sun is part of a larger community called a galaxy. A *galaxy* is a system of stars, dust, and gases all held together as a group by gravity. Our own galaxy is the Milky Way. It is just one of billions of other galaxies spread throughout the universe. The smallest known galaxies may contain only 100,000 stars, but the largest known galaxy, M87, contains over three billion stars.

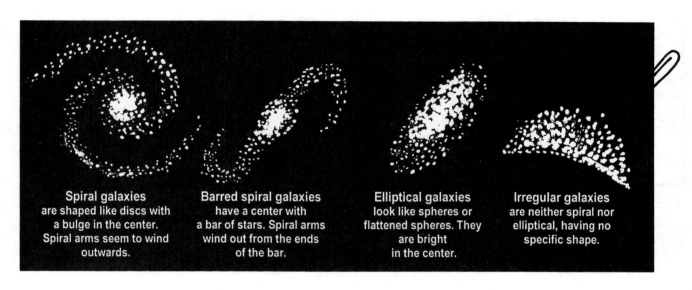

Spiral galaxies are shaped like discs with a bulge in the center. Spiral arms seem to wind outwards.

Barred spiral galaxies have a center with a bar of stars. Spiral arms wind out from the ends of the bar.

Elliptical galaxies look like spheres or flattened spheres. They are bright in the center.

Irregular galaxies are neither spiral nor elliptical, having no specific shape.

The Milky Way
(as it would look when viewed from the side)

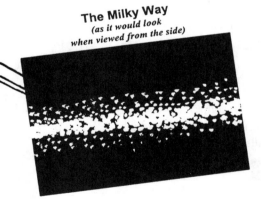

Our galaxy, the Milky Way, is a spiral galaxy with over 200 billion stars and plenty of dust and gas in between the stars.

It would take 130,000 light years to travel across the Milky Way.

Galaxies exists in groups, called ***clusters***. The Milky Way belongs to a cluster of 32 galaxies called the Local Group. The Milky Way is the second largest galaxy in the Local Group. Andromeda is the largest galaxy in this cluster. Clusters of galaxies seem to form larger clusters, called ***superclusters***. The Milky Way's Local Supercluster contains thousands of galaxies spread across over a million light-years of distance in space.

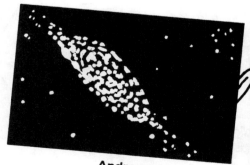

Andromeda

Get Sharp: The Universe

Taken together, all these galaxies, clusters, and superclusters, form the universe. The *universe* consists of everything that exists in space. It includes the Earth, the Sun, and everything in the solar system. In addition, it includes every other star and any planets it might have, all other celestial bodies (such as comets and asteroids), energy particles, magnetic fields, gases, and cosmic dust. Scientists who focus on questions about the universe are in the field of cosmology. *Cosmology* is the study of how the universe began, how it is now, how it's changing, and what the future might be.

What is a quasar?

Some distant galaxies have brilliant (though mysterious) starlike objects at their cores. These objects are called *quasars*, which is short for *quasi-stellar radio sources*. Some activity, probably due to the presence of a massive black hole at the center of a galaxy, causes this star-like object to emit enormous amounts of energy. This energy takes the form of light, radio waves, or other waves. These mysterious energy sources were first identified in 1963.

What is the Big Bang?

It is a widely-held scientific theory about the origin of the universe. According to the theory, the universe began a long time ago as one dense atom. Then a big, hot explosion occurred—a series of nuclear reactions that caused that small amount of matter to blow apart. The term **Big Bang** was coined in 1950 by British scientist Fred Hoyle.

Many scientists believe that, after the Big Bang, the matter in the universe has continued to fly apart. Will the universe fly apart forever?

What is the Big Squeeze?

The Big Squeeze is the opposite idea from the idea that the universe will continue to expand. Other scientists wonder if the universe will become so dense that the expansion will reverse. Then, gravitational forces might begin to pull everything back together again in a Big Squeeze.

Better Grades & Higher Test Scores / SCIENCE
Copyright ©2003 by Incentive Publications, Inc., Nashville, TN.

Get Sharp: The Universe

Stars

Stars are hot, bright, burning spheres of gas. From Earth, all stars may look like similar little spots of shimmering light. Stars may look as if they are splattered across a flat surface above us, all about the same distance from Earth. This is definitely not the case! Stars vary greatly in size, brilliance, luminosity, color, temperature, age, and distance from Earth.

Star Distances

Stars are different distances from Earth and from each other. Two stars that look as if they are next to each other may be millions of miles apart. The closest star to our solar system, Proxima Centauri, is 4.2 light years away. When you view a star in the sky, you are not seeing the star as it is now, but as it was when the light first left the star to travel across space.

Star Color

The temperature of a star affects its color. The hottest-burning stars are blue, followed by white, yellow, and red. During the main part of their lives (main sequence), stars are classified by their surface temperatures in this way:

Blue *(temperature about 60,000 Kelvins or K)*
White *(temperature about 9000 K)*
Yellow *(temperature about 6000 K)*
Red *(temperature about 2500-4000 K)*

Star Size

There is a great difference in the masses of stars from the time they are born. In star-talk, you might hear stars of different sizes referred to as *dwarfs, medium-sized stars* (like our Sun), *giants,* or *supergiants.*

Star Luminosity

Luminosity is the rate at which a star pours out energy. It is related to the size and surface temperature of a star. Luminosity and temperature are measures used by astronomers to classify stars. Most of the known stars are less luminous than our Sun.

Star Brilliance (or Magnitude)

The brightness of a star as seen from Earth is called its *apparent magnitude*. But since stars are not all the same distance from Earth, the appearance from Earth is not a true measure of brightness. A star's *absolute magnitude* is the measure of its actual brightness, determined as if it were a specific distance away (32.6 light years). Stars are given a magnitude number as a measure of brilliance. The brighter the magnitude, the lower the number will be. Magnitudes of some extremely bright stars even fall below zero.

Sirius, the brightest star in the sky, has a magnitude of -1.4 The faintest stars visible without a telescope are about Mag. 6.

We might think of stars as unchanging and everlasting. But they age, just like we do. Though it takes millions, billions, or trillions of years, a star has a whole life cycle—a birth, a youth, a middle age, and eventually, a death.

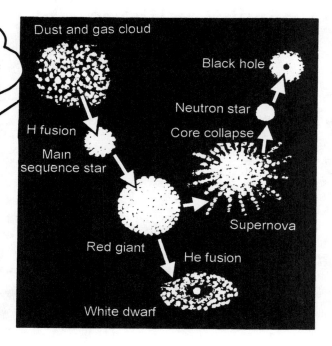

Dust and gas cloud

Black hole

H fusion

Main sequence star

Neutron star

Core collapse

Supernova

Red giant

He fusion

White dwarf

The Life Story of a Star

1 Stars are born in **nebulae**, vast clouds of dust and gas in space. Within a nebula, blobs of dust grow and become denser. As the density and gravity increase, the temperature climbs until it ignites nuclear reactions. The reactions cause the nebula to break up into a cluster of many baby stars.

2 After the star forms, it is in its main life period, called the **main sequence period**. A main sequence star lives and shines fairly steadily for millions of years or more.

3 Eventually, a star begins to burn itself out. It runs out of fuel and starts cooling down. As it cools, it collapses, causing temperatures to rise. The intense heat causes the gases to explode, so the star swells up into a glowing **red giant** that may be a hundred times the size of the original star.

4 From the red giant stage, a dwarf or medium-sized star (such as our Sun) slowly cools off and shrinks. It becomes a faint, small star called a **white dwarf**. Eventually it will fade out altogether into a **black dwarf**.

From the red giant stage, a giant or supergiant will blow up into a huge explosion called a **supernova**. A supernova may leave behind a tiny, dense, fast-spinning star called a **neutron star.** Such a star may give out radio waves in pulses as it rotates. These bursts of radiation are called **pulsars**.

5 There is some belief that a neutron star can shrink to a body so dense that the star disappears inside itself. This is known as a **black hole**. The idea is that the gravitational pull is so strong that everything nearby is sucked inside. Even light cannot escape.

Stars with greater mass are generally more luminous.

Dimmer stars have shorter lives.

Stars with greater mass have shorter lives.

Stars with greater mass have hotter temperatures.

In general, the hotter the star, the greater is its brilliance.

Get Sharp: Stars

Which Star is Which?

White dwarves, red giants, novas, double stars, neutron stars, quasars—so many stars! Here's a quick review to help you remember which stars are which.

protostar

A protostar is the first step in the evolution of a star. This is formed when nuclear fusion takes place in the core of the nebula, causing gas to begin glowing.

giant star

This is the label given to large stars, much larger than the Sun. Some stars are born as giants. Others only become giants near the end of their lives. (See *red giant*.)

red dwarves

The most common type of star in the universe is the red dwarf. Their small size results in cool temperatures, which leads to the red color. Most of them don't look red because they are too far away for the color to show.

main sequence star

This period in a star's life begins when the mass of the new star is stabilized. This life period that lasts for millions or billions (or even trillions) of years.

red giant

When a star begins to cool after the main sequence period, it expands outward and puffs up into a giant star with a red glow. All stars go through a red giant phase before they die.

dwarf stars

This is the label given to the smaller stars in the universe These stars are smaller than our Sun.

Polaris, the North Star, is a variable star whose brightness changes every four days.

blue giant

A blue giant is a star that is large and hot enough to burn at a very high temperature. A blue giant burns its fuel quickly with a very bright light. It has a short life.

medium stars

(also known as yellow stars)

Medium stars are similar in size to our Sun. They burn yellow because they have a medium temperature.

supergiant

Some stars are even bigger than giants, and are called supergiants. A massive star may become a supergiant as it gets older and begins to burn helium. This fuel burns hotter, so the star bulges out farther than normal.

A nebula

is a low-density cloud of gas and dust from which stars are born.

94

white dwarf

The dense white dwarf stage occurs after the red giant phase in small and medium stars. The star's matter collapses inward due to the pull of gravity and the star becomes extremely dense. It shines for a while with a dim white light.

supernova

A supernova is powerful explosion of a star that lights up the sky with tremendous heat and light. A supernova results from the collapse of a star that has massive size (larger than our Sun).

black hole

This is an extremely dense object that forms when a massive star collapses inward due to an immense gravitational force. When a giant star dies, it explodes in a supernova. What is left of the star gets very small, with so much gravitational pull that all the star's matter is pulled inside itself.

quasar

There are some extremely bright star-like objects at the very edges of the universe. These are objects about the size of our solar system that shine brighter than most galaxies because they produce so much light and energy.

variable star

Some stars change in their brightness, or appear to change in their brightness. This may be because the star expands and contracts. It may also occur because two or more stars orbiting a common point eclipse each other and the light is blocked out temporarily.

black dwarf

This is the phase when a star has totally burned out so that it no longer shines. It is the final phase in the life of a small or medium star.

neutron star

The core of a massive star ends up as a neutron star after a supernova. A neutron star is a rapidly-spinning star that gives off radio waves.

pulsar

Some neutron stars give off radio waves in a pulsing rhythm. These are called pulsars.

binary stars

This is a pair of stars that orbit each other or appear to orbit each other.

When in Doubt – Phone a Friend

Abigail, do you know what would happen if I got too close to a black hole?

It wouldn't be pretty! First a tremendous force would suck you in. Then, you'd get stretched out like a string of spaghetti. You would never get out. Even light cannot escape a black hole.

The bottom line is, Zelda, stay away from black holes!

Patterns in the Sky

Long ago, early stargazers invented *constellations*. They looked into the skies and imagined shapes or patterns created by groups of stars. Names of animals, ancient gods, people, or other creatures were given to these groups of stars called constellations. These constellations, plus some new ones, are still used today by star watchers.

Each constellation has a Latin name, but most are better known by a common English name. The sky contains 88 different constellations. The largest one is Hydra (the Water Snake) and the smallest one is Crux (the Cross). Often, a constellation is recognized by identifying its brightest star or stars. In the constellations shown here, the brightest stars are named.

The Northern Sky in January

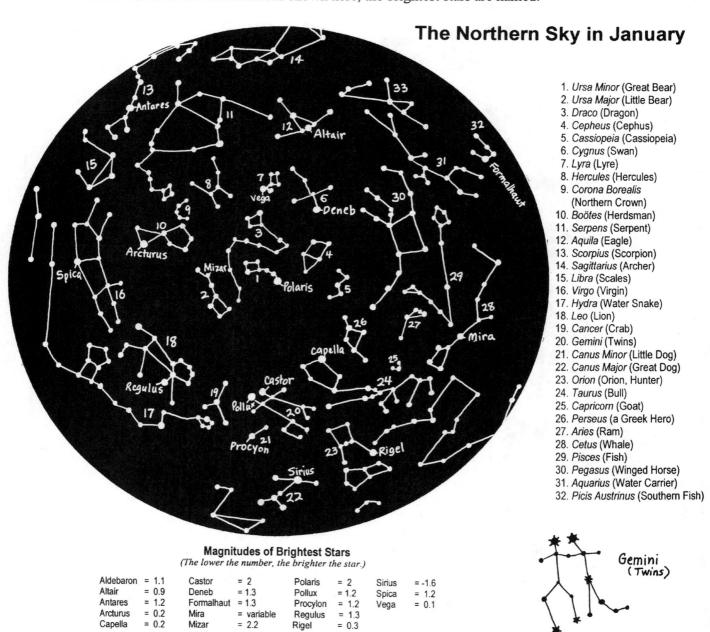

1. *Ursa Minor* (Great Bear)
2. *Ursa Major* (Little Bear)
3. *Draco* (Dragon)
4. *Cepheus* (Cephus)
5. *Cassiopeia* (Cassiopeia)
6. *Cygnus* (Swan)
7. *Lyra* (Lyre)
8. *Hercules* (Hercules)
9. *Corona Borealis* (Northern Crown)
10. *Boötes* (Herdsman)
11. *Serpens* (Serpent)
12. *Aquila* (Eagle)
13. *Scorpius* (Scorpion)
14. *Sagittarius* (Archer)
15. *Libra* (Scales)
16. *Virgo* (Virgin)
17. *Hydra* (Water Snake)
18. *Leo* (Lion)
19. *Cancer* (Crab)
20. *Gemini* (Twins)
21. *Canus Minor* (Little Dog)
22. *Canus Major* (Great Dog)
23. *Orion* (Orion, Hunter)
24. *Taurus* (Bull)
25. *Capricorn* (Goat)
26. *Perseus* (a Greek Hero)
27. *Aries* (Ram)
28. *Cetus* (Whale)
29. *Pisces* (Fish)
30. *Pegasus* (Winged Horse)
31. *Aquarius* (Water Carrier)
32. *Picis Austrinus* (Southern Fish)

Magnitudes of Brightest Stars
(The lower the number, the brighter the star.)

Aldebaron	= 1.1	Castor	= 2	Polaris	= 2	Sirius	= -1.6
Altair	= 0.9	Deneb	= 1.3	Pollux	= 1.2	Spica	= 1.2
Antares	= 1.2	Formalhaut	= 1.3	Procylon	= 1.2	Vega	= 0.1
Arcturus	= 0.2	Mira	= variable	Regulus	= 1.3		
Capella	= 0.2	Mizar	= 2.2	Rigel	= 0.3		

Gemini (Twins)

The Southern Sky in January

1. *Crux* (Cross)
2. *Centaurus* (Centaur)
3. *Lepus* (Hare)
4. *Ara* (Altar)
5. *Triangulum Australe*
6. *Octans* (Octant)
7. *Libra* (Scales)
8. *Scorpio* (Scorpion)
9. *Sagittarius* (Archer)
10. *Ophiuchus* (Serpent Holder)
11. *Hercules* (Hercules)
12. *Corona Australis*
    (Southern Crown)
13. *Boötes* (Herdsman)
14. *Virgo* (Virgin)
15. *Hydra* (Water Snake)
16. *Leo* (Lion)
17. *Cancer* (Crab)
18. *Carina* (Keel)
19. *Canis Major* (Great Dog)
20. *Canis Minor* (Little Dog)
21. *Gemini* (Twins)
22. *Orion* (Hunter)
23. *Taurus* (Bull)
24. *Eridanus* (River Eridanus)
25. *Lupus* (Wolf)
26. *Aries* (Ram)
27. *Cetus* (Whale)
28. *Pices* (Fish)
29. *Phoenix* (Phoenix)
30. *Grus* (Goat)
31. *Aquarius* (Water Carrier)
32. *Pegasus* (Winged Horse)
33. *Aquila* (Eagle)
34. *Delphinus* (Dolphin)
35. *Cygnus* (Swan)

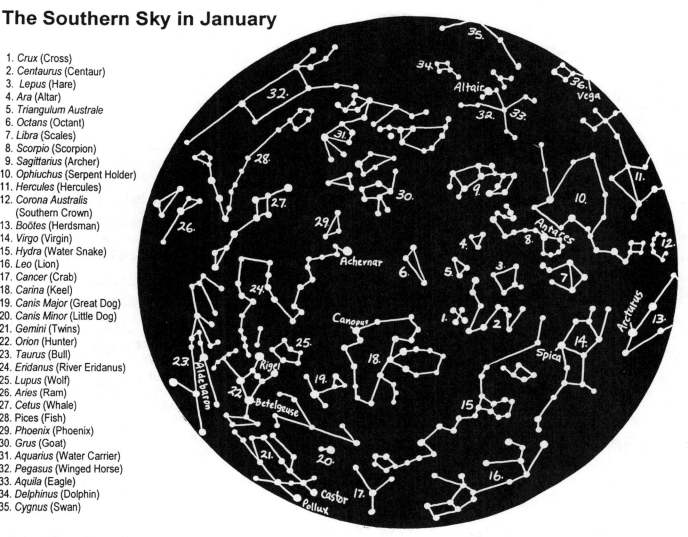

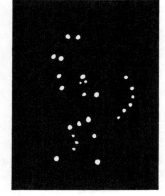

Magnitudes of Brightest Stars
(The lower the number, the brighter the star)

Achnernar	= 0.6	Antares	= 1.2	Canopus	= -0.9	Rigel	= 0.3
Aldebaron	= 1.1	Arcturus	= 0.2	Castor	= 2	Spica	= 1.2
Altair	= 0.9	Betelgeuse	= 0.9	Pollux	= 1.2	Vega	= 0.1

Orion, the Hunter

Leo, the Lion

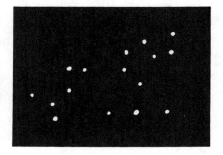

Exploring the Skies

As you can imagine, the skies have been a source of wonder and mystery for people from the beginning of life on Earth. Human curiosity has turned eyes towards space for centuries.

A Timeline of Some Major Events & Discoveries

B.C.

4800   The Babylonians kept astronomical records.

335   Aristotle showed that the Earth is round.

300s   Aristotle developed a system of astronomy.

240   The comet that later became known as Halley's Comet was first sighted.

A.D.

140   Ptolemy put forth his theory that the Earth was the center of the universe.

1543   Copernicus claimed that the Sun, not the Earth, was at the center of the universe.

1600s   Kepler showed that planets move in elliptical orbits.

1608   The first telescope made by Dutchman Hans Lippershey.

1609   Galileo examined the sky with a telescope.

1671   The reflecting telescope invented by Newton.

1675   The Royal Greenwich Observatory was founded in England.

1781   Uranus was discovered. (It was the first planet discovered with a telescope).

1846   German astronomer Johann G. Galle discovered the planet Neptune.

1914   Eddington identified spiral galaxies for the first time.

1915   The Theory of Relativity was published by Einstein

1920s   Edwin Hubble demonstrated that the universe is expanding.

1926   American Robert Goddard launched the first liquid-propellant rocket.

1930s   Physicist Hans Bethe explained how nuclear fusion powers stars.

1930   American astronomer Clyde W. Tombaugh discovered the planet Pluto.

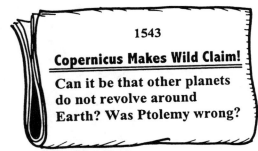

1543

Copernicus Makes Wild Claim!

Can it be that other planets do not revolve around Earth? Was Ptolemy wrong?

1609

New Astronomy Tool

A long tube allows people to get a better look at the stars. Galileo has been making amazing discoveries with Lippershey's new invention.

1925

Hubble Says Universe is Expanding

Will the whole universe eventually fly apart?

1930

Another Planet!

A ninth planet has been discovered. Some scientists think it's a wandering moon, but Pluto has been officially declared a planet.

98

Pets in Space?

A Space Race First
November 3, 1957

A dog named Laika was the first living creature to travel into space. She was launched by the Soviet Union aboard *Sputnik 2*. The mission did not include a way to return the dog to Earth. So Laika died after a week in space.

MONKEYING AROUND in SPACE
1958

A monkey named Gordo was launched into space aboard a *Jupiter* flight. Scientists wanted to explore the safety of space flights for living creatures. Gordo suffered no adverse effects from the space trip, but a failure in the rocket prevented his recovery.

Animals Survive Space Trip
1959

Two space monkeys, Able and Baker, made history as the first living beings to be successfully recovered after a space flight. They were launched on a *Jupiter* mission and traveled 1500 miles in space.

Zoo Blasts Off
1973

Two spiders, Anita and Arabella, took a trip on board the Space Lab. They were not alone! 720 fruit flies, 6 mice, 2 minnows and many minnow eggs accompanied them. Other space flights have taken bullfrogs, hamsters, rats, dogs, cats, monkeys, and chimps.

1957 The U.S.S.R. launched *Sputnik 1*, the first artificial satellite.

1957 The U.S.S.R. launched *Sputnik 2*, the first satellite carrying a living being (a dog) into space.

1958 NASA *(The National Aeronautics and Space Administration)* was formed.

1958 *Explorer 1* was the first satellite launched by the United States.

1959 The U.S.S.R. launched *Luna 2*, the first space probe to orbit the Moon.

1961 Soviet cosmonaut Yuri A. Gagarin was the first person to orbit the Earth.

1961 Alan B. Shepard, Jr., was the first American launched into space.

1962 John H. Glenn, Jr., was the first American to orbit the Earth.

1962 *Ranger 4* was the first U.S. space probe to hit the Moon.

1962 The U.S. *Mariner 2* space probe visited Venus.

1962 *Telstar 1*, the first communications satellite, was launched.

1963 Soviet cosmonaut Valentina Tereshkova became the first woman in space.

1963 Dutch astronomer Maarten Schmidt identified *quasars*.

1964 The Soviet Union launched *Voskhod,* the first multi-person space capsule.

1965 Alexsei Leonov, traveling in *Voshkod 2*, performed the first space walk.

1967 Three *Apollo* astronauts were killed in a fire that occurred on the launch pad before their rocket launched.

1967 British astronomer Jocelyn Bell Burnell identified *pulsars*.

1968 The U.S. launched *Apollo 8,* the first manned spacecraft to orbit the Moon.

1969 On July 20, U.S. astronauts Neil A. Armstrong and Buzz Aldrin became the first humans to land on the Moon.

1971 The NASA space probe, *Mariner 9*, was the first spacecraft to orbit Mars.

1971 The Soviet space probe, *Mars 3*, was the first craft to land on Mars.

1975 Soviet and U.S. spacecraft linked in space as part of the first international space mission.

1975 U.S. probe *Viking 1* was launched. Along with *Viking 2*, it sent photos and data back from the surface of Mars.

1977 U.S. probe *Voyager 2*, was launched. Over the next several years, this probe flew past and photographed Jupiter, Saturn, Uranus, and Neptune.

Better Grades & Higher Test Scores / SCIENCE
Copyright ©2003 by Incentive Publications, Inc., Nashville, TN.

Get Sharp: Space Exploration

1981 U.S. launched the space shuttle *Columbia*, the first reusable piloted spacecraft.

1983 U.S. launched *Pioneer 10* which traveled beyond all the planets.

1985 U.S. space shuttle *Challenger* exploded after launch, killing all seven crew members.

1986 The first module of the *Mir Space Station* was launched.

1988 NASA resumed the U.S. space program with the launch of the shuttle, *Discovery*

1989 The U.S. launched the probe *Galileo*, which reached Jupiter in 1995.

1990 U.S. space probe *Magellan* began its orbit of Venus.

1990 The *Hubble Space Telescope* was launched.

1990-94 U.S. space probe *Magellan* mapped the surface of Venus.

1994 Comet Shoemaker-Levy collided with Jupiter.

1996 The U.S. launched the *Mars Global Surveyor* to map the planet Mars.

1996 *Mars Pathfinder* sent important data about the planet back to Earth.

1997 The U.S. launched the probe *Cassini*, due to reach Saturn in 2004.

1998 The *Mars Climate Orbiter*, a weather satellite headed for Mars, was lost when it arrived on the planet.

1998 The first part of the *International Space Station* was launched into place.

1999 The *Mars Polar Lander/Deep Space 2* spacecraft, which was to set down near Mars' south polar cap, was lost during landing.

2001 The first full-time crew (one American astronaut and one Russian cosmonaut) occupied the *International Space Station*.

2001 Space tourist Dennis Tito paid for and took a trip to the *International Space Station*.

2001 The *Mir Space Station* was destroyed (purposely).

2001 *Quaoar*, a large celestial body beyond Pluto, was discovered. It is the largest body found since the 1930 identification of Pluto.

Bulletin!

Tourists in Space: A Good Idea?

Dennis Tito was the first space tourist. But will he be the last? There is increased interest among adventurers who have enough money to buy their way into outer space. Is this a blossoming new industry? Should tourists go into space? What are the dangers and advantages of this practice?

Learn a lot more about the solar system and outer space from the NASA websites. The information is fun and fascinating!
www.nasa.gov
www.kids.msfc.nasa.gov
And don't pass up a change to learn about the amazing discoveries of the Hubble Space Telescope. Visit the Hubble Site.
www.hubblesite.org

Better Grades & Higher Test Scores / SCIENCE
Copyright ©2003 by Incentive Publications, Inc., Nashville, TN.

GET SHARP →

on Earth Science

Ah, Earth!
The sun, the sand, the sea,
the gentle breezes...

...the food!

The Earth's Structure

Viewed from space, Earth looks like a sphere covered with clouds and water. Closer up, an observer can see large masses of land surrounded by water. Basically, Earth is a huge ball of rock, covered with water and land, and wrapped in an envelope of air.

Layers

One way to describe Earth's structure is to tell about each of its layers, or *spheres*.

The atmosphere: a blanket of air that surrounds Earth, about 300 miles thick, and made mostly of gases

The lithosphere: the outer rocky region that includes Earth's crust and the rigid upper layer of the mantle

The hydrosphere: the water (liquid and frozen) found on Earth's surface

The biosphere: the regions of the land, water, and atmosphere inhabited by living things

The asthenosphere: the part of the mantle that is almost fluid, having a plastic-like consistency

Earth

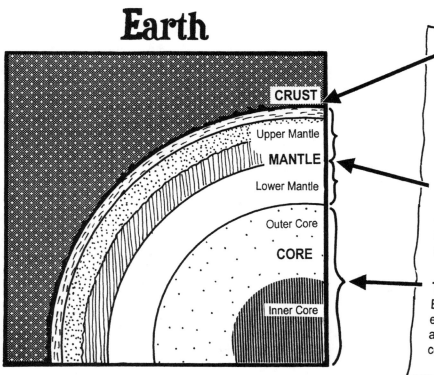

The crust is a relatively thin layer of rock, ranging from 3-31 miles thick. The crust is thinnest beneath the oceans. (This is oceanic crust.) Crust beneath the continents (continental crust) is thicker. Earth's crust is thickest beneath mountain ranges.

The mantle is a layer of hot rock (magma) about 1800 miles thick. The uppermost part is rigid. The lower mantle region is a material with the consistency of a very thick liquid.

The core is the center region of Earth, about 2200 miles thick. It is extremely hot rock that is mostly iron and nickel. The innermost part of the core is solid; the outer core is liquid.

Earth's Vital Statistics

Age - about 4.5 billion years
Mass - 6.6 sextillion tons
Polar circumference - 24,850 miles
Equatorial circumference - 24,902 miles
Total surface area - 196,900,000 miles
Land Area - 57,100,000 miles
Water Area - 139,800,000 miles

Highest land point - Mt. Everest, 29,035 feet
Lowest land point - Dead Sea, 1310 feet below sea level
Core temperatures - 5000° to 9000° F
Surface temperature - -129° to 136° F
Lowest point - Mariana Trench, Pacific Ocean, 25,840 feet below sea level
Distance to the center - about 4000 miles

The Earth's surface is 71% water! We should wear our swim suits 71% of the time!

Soil

Most land surfaces on Earth are covered with soil. Soil may be just a thin layer on top of rock, or it may be over 100 feet thick. *Soil* is a mixture of decayed organic material, weathered rock, air, and water. Soil is formed when forces such as wind, ice, or running water break down rocks and other materials. This process takes hundreds of years. Different soil types have different colors, textures, and chemical make-ups. Soil can be waterlogged or well-drained. It can be rich or poor in organic matter.

THE DIRT ON DIRT

Most soil is about 50% rocks and minerals.

Residual soil is soil that forms from the weathering of rock in the place it is found.

Transported soil is soil that has been formed somewhere else. It is eroded from one place and moved to another.

Soil particles are the organic and mineral particles in soil.

Pore spaces are spaces between soil particles.

Soil horizons are layers that develop in soil.

Humus is decomposed living matter in soil.

Topsoil is the highest horizon in soil.

Subsoil is the lowest horizon in soil.

Bedrock is the solid rock beneath soil.

Leaching is the process by which components of soil are dissolved in water and carried downwards.

Clay is soil of fine particles less than 0.001 in. thick

Silt is soil of fine particles, 0.001-0.002 in. thick.

Sand is soil of coarser particles, .002-.08 in. thick

Loam is a soil that is a mixture of clay, silt, sand, and organic material.

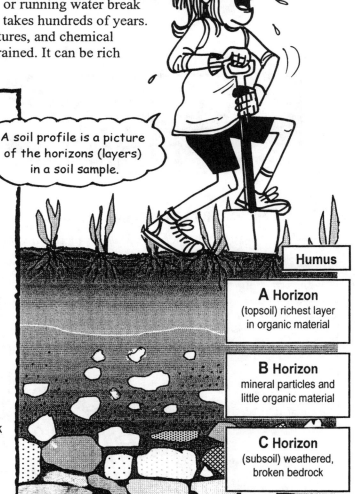

A soil profile is a picture of the horizons (layers) in a soil sample.

Humus

A Horizon
(topsoil) richest layer in organic material

B Horizon
mineral particles and little organic material

C Horizon
(subsoil) weathered, broken bedrock

Bedrock

Internal Processes

The Earth is always changing. Many of the changes begin deep inside the Earth. One theory explains many of these changes.

Plate Tectonics

Get Sharp Tip # 5

The word *tectonics* comes from a Greek word which means *builder*.

The **Theory of Plate Tectonics** is a current scientific explanation for the movements within Earth. It explains processes such as earthquakes, volcanoes, mountain building, and continental drift. This theory suggests that Earth's crust, like a giant jigsaw puzzle, is made up of huge rigid plates (*the lithosphere*) that lie or "float" on the plastic-like lower layer of the mantle (*the asthenosphere*). It is believed that these plates move constantly, usually at a slow pace. They slide past each other, slide under or over one another, push together, scrape against each other, and pull apart. The energy generated by movement of the plates causes many of the changes or activity we see on Earth's surface.

INTERNAL FORCES

Compression	Tension	Shearing
forces that push against a solid substance directly and evenly from opposite sides and squeeze it until it folds	force that pulls a solid apart or stretches it	forces that are not directly opposite pushing against a solid from different sides, resulting in tearing and twisting

THE ACTION IS AT THE EDGES!

Most of the activity (such as volcanoes or earthquakes) happens at the edges, or boundaries, of the plates.

Divergent boundaries are boundaries where two plates pull apart.

Convergent boundaries are boundaries between two colliding plates. When two oceanic plates collide (or one oceanic and one continental), one is forced down into the asthenosphere. The area where the plate submerges is the **subduction zone**. When two continental plates collide, the rocks crumple and rise, forming folded mountains.

Transform faults are boundaries where plates slide or scrape past each other. These are often areas of earthquake activity.

These are some processes and changes resulting from the movements beneath Earth's surface.

Continental Drift

There is a belief that the continents were once connected in a *supercontinent* sometimes called *Pangaea*. The *Continental Drift Theory* suggests that the continents moved apart (leaving ocean basins between them), and that they are still moving.

Seafloor Spreading

On the Atlantic Ocean, the ridge of the Mid-Atlantic Mountain Range is on the boundary of two plates. As the plates move away from a crack in the crust, melted magma flows into the crack and hardens to form new seafloor. This process is called *seafloor spreading*.

Folds, Faults, & Fractures
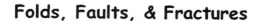

Fractures are breaks in rock that result from some sort of force beneath Earth's surface.

Folds are bends in the rock. When rock is more elastic, it may bend into folds that alternate ridges (*anticlines*) with troughs (*synclines*).

Faults are fractures along which some movement takes place. Movement can occur in any direction along the fault. Faults occur in areas where Earth's rocky crust is weak.

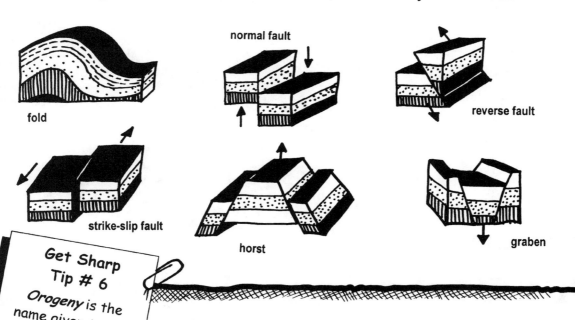

fold

normal fault

reverse fault

strike-slip fault

horst

graben

Get Sharp Tip # 6
Orogeny is the name given to the processes that build mountains. These include volcanic activity, folding, crumpling, and upwarping.

Mountain Building

Mountains are formed when molten rock from inside the Earth pushes a dome in the surface, or when it pours out of the surface to form a cone. Mountains are also formed when internal forces thrust some of Earth's crust upwards or cause the crust to fold.

Earthquakes & Volcanoes (See pages 106-109.)

105

Earthquakes

Earthquakes are among the most frightening and damaging of the natural processes. An earthquake occurs when rocks move on opposite sides of a *fault*, or break, in the crust. The actual quake is the shaking caused by the vibrations resulting from the break.

How an Earthquake Travels

When rock breaks within Earth's lithosphere, it releases energy in the form of waves that travel out from the fault in all directions. These are called *seismic waves*.

The Primary Waves (**P** *waves,* or *compressional waves*) are underground. These travel fast through the earth. They move back and forth in the direction the wave is traveling.

Secondary waves (**S** *waves,* or *shear waves*) are also underground waves. These move in a side-to-side motion, traveling more slowly than the primary waves.

Surface waves travel along the surface in a rolling motion.

> As many as 8000 earthquakes might occur each day. Most of those are not strong enough to be felt.

> ## Get Sharp Tip # 7
>
> The hypocenter of an earthquake is the center of the earthquake deep within the Earth—the source where the shock waves originate. The epicenter is the point on Earth's surface above the hypocenter. The most violent shaking is felt near the epicenter.

Studying & Measuring Earthquakes

Scientists who study earthquakes are called *seismologists*. They use machines to determine the size of a quake. The *magnitude* is a measure of the energy released by the earthquake at its source. The magnitude of a quake will be the same no matter where a person is in relation to the epicenter. The *intensity* of a quake is a measure of the amount of shaking and damage. This measure differs by location.

Richter Magnitude

The Richter Scale is commonly used to compare the sizes of earthquakes. The scale is a mathematical formula that determines magnitude by recording the amplitude of seismic waves during a certain period of time. A machine called a *seismograph* detects earthquakes and measures the magnitudes. It records a zig-zag line that shows the strength of the ground movements.

Moment Magnitude

The *Moment Magnitude Scale* is another way to find earthquake size. It measures large earthquakes more precisely than the Richter Scale because it measures more of the ground movements.

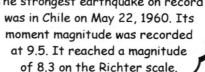

> The strongest earthquake on record was in Chile on May 22, 1960. Its moment magnitude was recorded at 9.5. It reached a magnitude of 8.3 on the Richter scale.

Earthquake Damage

Earthquakes can lead to tremendous loss of life and destruction to property. An earthquake can cause damage in many ways. The ground may crack. Buildings, bridges, dams, and other structures may collapse. Broken glass, falling objects, and landslides can cause injury or damage. Huge ocean waves (tsunamis) created by the vibrations can cause floods. Fallen power lines and cracked gas pipes can cause fires. Hazardous chemicals may be spilled during earthquakes or sewage may seep into water supplies from broken sewer lines.

Aftershocks

Small earthquakes often follow a large earthquake. These "follow-up" tremors are called *aftershocks*. There are usually several of these following a major quake. Sometimes the aftershocks can be almost as strong and can cause as much damage as the original quake.

The deadliest earthquake in modern times was in Tang-shan, China in 1976. About 250,000 people were killed.

The fires that erupted after the San Francisco Earthquake of 1906 took many lives and destroyed a great portion of the city. It is believed that more people died from the fires after the quake than died from the earthquake itself.

According to the *Guinness Book of Records,* the largest tsunami rose 1720 feet in Lituya Bay, Alaska, in 1958.

What to do in an earthquake...

- DROP, COVER, AND HOLD ON!
- Crawl to a safe place under a strong table or in a doorway away from the outside of the house and away from windows or heavy objects that can fall on you.
- Stay indoors until the shaking stops and you're sure it's safe to exit.
- Stay away from windows.
- If you are outdoors, drop to the ground. Try to crawl or scurry to a clear spot away from power lines, buildings, and trees.
- When the shaking stops, take care of any injuries or call for help if injuries are serious.
- Extinguish small fires, and turn off any gas.
- Be ready for aftershocks. When you feel one, DROP, COVER, AND HOLD ON.

HELP!

For more fascinating information about earthquakes, visit the U.S. Geological Survey earthquake site for kids: http://earthquake.usgs.gov/4kids/

Volcanoes

A volcano is another one of those sudden, dramatic, violent, and dangerous processes of nature. A volcano is any natural opening in Earth's crust through which hot melted rock (magma), fragments of rock, ash, and hot gases erupt and pour out over the surrounding land. Volcanoes can be very destructive to life and land. Volcanoes have killed thousands of people and buried whole villages. A volcanic eruption can also trigger a tsunami, causing even more death and devastation.

A volcano is caused by pressure within the Earth forcing melted rock to the surface. The molten rock with dissolved gases lies in a **magma chamber** deep beneath the opening in the crust. When pressure builds up, the magma and gases escape through a tube-like passage called a **vent**. Sometimes **secondary vents** develop farther down from the **main vent**. When the volcano erupts, it spews gas, dust, ash, and rocks into the air. When the magma reaches outside Earth's crust, it is called **lava**. The hot lava flows down the sides of the volcano. Sometimes, after an eruption, the top of a volcano collapses, forming a basin-like depression called a **crater**.

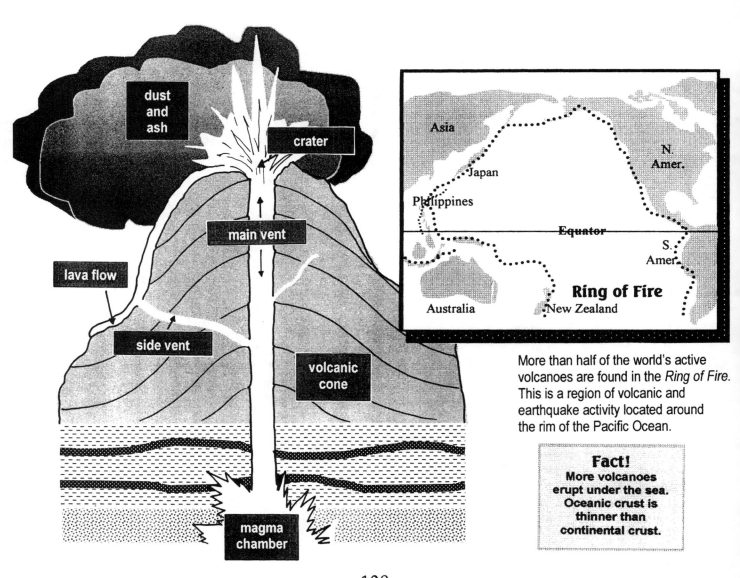

More than half of the world's active volcanoes are found in the *Ring of Fire*. This is a region of volcanic and earthquake activity located around the rim of the Pacific Ocean.

Fact!
More volcanoes erupt under the sea. Oceanic crust is thinner than continental crust.

Kinds of Volcanoes

Volcanoes are grouped according to their shape and the type of material. The shape of a volcano depends on the type of eruption that created it.

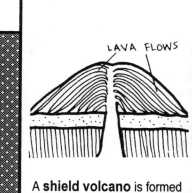

LAVA FLOWS

A **shield volcano** is formed from basalt which flows slowly out a central opening, resulting in a low, dome-shaped mountain.
Example: Mauna Loa and other volcanoes in the Hawaiian Islands

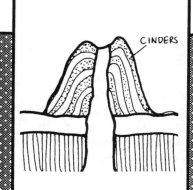

CINDERS

A **cinder cone volcano** is formed by violent eruptions that blow lava out in the form of cinders, forming a cone-shaped mountain.
Example: Mt. Pelée

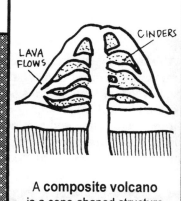

LAVA FLOWS   **CINDERS**

A **composite volcano** is a cone-shaped structure formed when flows of lava, ash, rock, and cinders erupt from a single vent.
Examples: Mt. Fiju, Mt. Vesuvius, Mt. St. Helens

An **extinct** volcano has been inactive for as long as any records have been kept.

A **dormant** volcano is inactive, but could become active again.

Some Memorable Eruptions

Mt. Vesuvius, Italy	79 and 1631
Etna, Sicily	1169 and 1669
Miya-Yama, Java	1793
Tambora, Indonesia	1815
Mauna Loa, Hawaii,	1872
Pelée, Martinique	1902
Vesuvius, Italy	1906
St. Helens, U.S.	1980
Nevado del Ruiz, Columbia	1985

For more fascinating volcano information, visit the US Geological Survey's volcano website:
http://volcanoes.usgs.gov

An **active** volcano erupts constantly.

An **intermittent** volcano erupts at regular intervals.

109

Minerals & Rocks

There are almost 100 naturally occurring elements in Earth's crust, but most of the crust is made up of just a few. The elements combine to form minerals. All rocks are made up of minerals. Together, rocks and minerals make up just about all of the Earth.

What did the teenage rock say to her boyfriend?

Minerals

A *mineral* is a naturally occurring, inorganic solid. Each mineral has a particular chemical make-up arranged in a crystalline pattern. There are over 2000 minerals that have been identified. A few minerals consist of a single element. The other elements are combinations of minerals. Almost all minerals are made from these elements: oxygen, silicon, aluminum, iron, magnesium, sodium, potassium, and calcium. A mineral generally contains the same kinds of atoms in the same proportions, arranged the same way. The structure of most minerals fits into one of these six crystal systems.

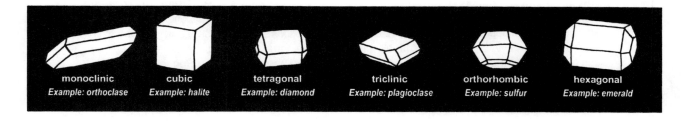

monoclinic	cubic	tetragonal	triclinic	orthorhombic	hexagonal
Example: orthoclase	*Example: halite*	*Example: diamond*	*Example: plagioclase*	*Example: sulfur*	*Example: emerald*

Gems

A *gemstone* is a mineral that is especially rare and stands apart from other minerals because of its beauty. Color, luster, radiance, and hardness are properties that make a mineral valuable as a gem. Diamonds, rubies, opals, emeralds, and sapphires are all gemstones.

I don't know! What did the teenage rock say to her boyfriend?

Identification of Minerals

Minerals are identified by their physical properties. You can learn to recognize many common minerals by paying attention to or testing these characteristics.

1. **Color**
2. **Form** is the shape or texture of the mineral (crystal shapes, grainy fibers).
3. **Specific gravity** is the mass of a mineral in relation to water.
4. **Hardness** is the mineral's resistance to being scratched.
   A harder mineral scratches a softer one.
5. **Luster** is the way light is reflected from the surface of the mineral.
   The luster can be *metallic* (shiny) or n*onmetallic* (dull, glassy or pearly).
6. **Streak** is the color a mineral makes when it is rubbed across of a piece of porcelain tile.
7. **Breakage** is the way a mineral breaks. *Cleavage* is breakage along smooth, flat planes.
   Fracture is irregular breakage (rough, jagged, or curved).
8. **Special characteristics** (tastes, odors, sounds, reactions with other substances)

I'm tired of being taken for granite.

Physical Properties of Some Common Minerals

Metallic Luster

Mineral	Color	Streak	Hardness	Crystals
GRAPHITE	black to gray	black to gray	1-2	hexagonal
SILVER	silvery, white	light gray to silver	2.5	cubic
GALENA	gray	gray to black	2.5	cubic
GOLD	pale golden-yellow	yellow	2.5-3	cubic
COPPER	copper red	copper red	3	cubic
CHROMITE	black or brown	brown to black	5.5	cubic
MAGNETITE	black	black	6	cubic
PYRITE	light brassy yellow	greenish black	6.5	cubic

The Cullinan diamond is a South African diamond found in 1905. At 3106 carats, it is the largest diamond ever found.

Nonmetallic Luster

Mineral	Color	Streak	Hardness	Crystals
TALC	white, greenish	white	1	monoclinic
GYPSUM	colorless, gray, white	white	2	monoclinic
SULFUR	yellow	yellow to white	2	orthorhombic
MUSCOVITE	white, gray, yellow, rose, green	colorless	2.5	basal cleavage
HALITE	colorless, red, white, blue	colorless	2.5	cubic
CALCITE	colorless, white	colorless, white	3	hexagonal
DOLOMITE	colorless, white, pink, green, gray	white	3.5-4	hexagonal
FLOURITE	colorless, white, blue, green, red, yellow, purple	colorless	4	cubic
HORNBLENDE	green to black	gray to white	5-6	monoclinic
FELDSPAR	gray, green, white	colorless	6	monoclinic
QUARTZ	colorless, many colors	colorless	7	hexagonal
GARNET	yellow-red, green, black	colorless	7.5	cubic
TOPAZ	white, pink, yellow, blue, colorless	colorless	8	orthorhombic
CORUNDUM	colorless, blue, brown	colorless	9	hexagonal
DIAMOND	colorless, many colors	colorless	10	octagonal or hexagonal

$9.8 million was the price paid for the most expensive diamond sale on record. The rough 225-carat gem was sold in 1989.

The beautiful green emerald is one of most rare and valuable of all gems. Emeralds are found in Brazil, Colombia, Siberia, Madagascar, and the U.S. Its composition is $Be_3Al_2(Si_6O_{18})$

Moh's Hardness Scale

Mineral	Hardness	Hardness Test
TALC	1	softest, can be scratched by a fingernail
GYPSUM	2	soft, can be scratched by a fingernail but cannot be scratched by talc
CALCITE	3	can be scratched by a penny
FLOURITE	4	can be scratched by a steel knife or a nail file
APATITE	5	can be scratched by a steel knife or nail file, but not easily
FELDSPAR	6	knife cannot scratch it; it can scratch glass
QUARTZ	7	scratches glass and steel
TOPAZ	8	can scratch quartz
CORUNDUM	9	can scratch topaz
DIAMOND	10	can scratch all others

Rocks

All rocks are formed from single minerals or combinations of minerals. The rocks of Earth fall into three main groups, and are classified according to the way they are formed.

Granite

Igneous Rocks

Igneous rocks are formed from hot, liquid Earth materials (magma or lava).

Intrusive igneous rocks form when magma trapped below the surface cools slowly. The slow cooling leaves intrusive rocks with coarse textures.
Examples: olivine, peridotite, diorite, granite, orthoclase, quartz, basalt

Extrusive igneous rocks form when lava flowing on the surface cools quickly. The fast cooling results in rocks that are smooth and glassy or finely grained.
Examples: felsite, andesite, rhyolite, obsidian, pumice

Metamorphic Rocks

Metamorphic rocks are rocks that have been changed by heat and pressure.

Foliated rocks have bands.
Examples: slate, phyllite, schist, gneiss

Nonfoliated rocks have no banding. They are usually massive.
Examples: quartzite, anthracite, graphite, marble

Marble

Sedimentary Rocks

Sedimentary rocks are made of sediments (loose material on Earth's surface). When weathered rocks break up and are deposited on the ground they eventually settle and harden in layers. Sedimentary rocks are made of series of layers.

Clastics are sedimentary rocks made from fragments of rocks, minerals, and shells.
Examples: sandstone, siltstone, shale, conglomerate, breccia

Nonclastics are sedimentary rocks formed from organic material or from chemical reactions.
Examples: calcite, limestone, peat, coal, flint, chalk, alabaster, rock gypsum

Shale

> Rocks are constantly changing. Each kind of rock can be changed into the other two kinds. The rock cycle shows changes undergone by rocks.

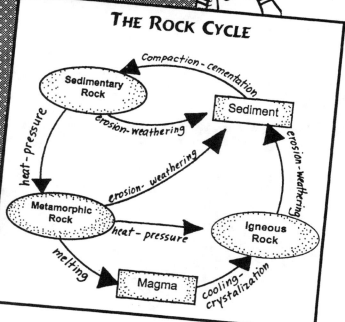

THE ROCK CYCLE

Earth History

Rocks are like clues in a huge mystery. Earth's history is the mystery. Rocks help scientists tell the story of Earth's past. Many tools are used to find out about the ages of Earth's rocks and to learn about the processes of the past. The layers of sedimentary rocks on Earth's surface are some of the best sources of clues.

Dating Earth's Rocks

Era	Period	Millions of years ago
Cenozoic	**Quaternary**	**1.8-present**
	Holocene (epoch)	0.1-present
	Pleistocene (epoch)	1.8-0.1
	Tertiary	**65-1.8**
	Pliocene (epoch)	5.3-1.8
	Miocene (epoch)	23.8-5.3
	Oligocene (epoch)	33.7-23.8
	Eocene (epoch)	55.5-33.7
	Paleocene (epoch)	65-55.5
Mesozoic	**Cretaceous**	**145-65**
	Jurassic	**213-145**
	Triassic	**248-213**
Paleozoic	**Permian**	**386-248**
	Pennsylvanian	**325-286**
	Mississippian	**360-325**
	Devonian	**410-360**
	Silurian	**440-410**
	Ordovician	**505-440**
	Cambrian	**544-505**

Precambrian Time
4500 – 540 million years ago

The **law of superposition** is a principle that helps geologists find the relative ages of rocks in a rock bed. This law supposes that rock is layered in order of age or time. This would mean that the lower layers are the oldest, and the upper layers are the youngest.

Scientists also use **fossil correlation** to help date the rocks. They try to reconstruct the history by looking at the types of fossils found in each layer.

Relative dating is a method used to put events in Earth's history in sequence in relationship to each other. Scientists theorize a geologic history of an area from the layers of rocks and their relationship to each other.

Absolute dating is a method that gives the actual time of an event in Earth's history. It is based on the amount of time it takes for certain elements in the rocks to decay.

A **geologic time scale** shows the sequence of events in Earth's history. The time scale is organized by time periods according to changes in life forms. The largest divisions are *eras.* Each era is divided into *periods* and periods are divided into *epochs.*

Is it **relative dating** when you take your second cousin to a **rock** concert?

Absolutely not!

Get Sharp: Earth History

Wonders of Earth's Surface

The surface of Earth is a wonderful collection of peaks and troughs, dry lands and wild seas, breathtaking panoramas and curious formations. Here's a brief, illustrated glossary that will give you a review and overview of some of Earth's major features. Match the numbers to see many of the features on the map (page 115).

1. **archipelago** – large group or chain of islands
2. **fjord** – deep narrow inlet of the sea between steep cliffs
3. **peak** – pointed top of a mountain or hill
4. **mountain** – high, rounded or pointed landform with steep sides
5. **mountain range** – row or chain of mountains
6. **foothills** – hilly area at the base of a mountain range
7. **hill** – rounded, raised landform, lower than a mountain
8. **cliff** – steep rock face
9. **volcano** – mountain created by volcanic action
10. **strait** – narrow waterway or channel connecting two bodies of water
11. **canyon** – deep, narrow valley with steep sides
12. **ocean** – one of Earth's four largest bodies of water
13. **waterfall** – flow of water falling from a high place to a low place
14. **glacier** – large sheet of ice that moves slowly over a land surface
15. **bay** – part of a large body of water that extends into the land
16. **lake** – body of water completely surrounded by land
17. **cape** – part of a coastline that projects into the water
18. **gulf** – part of an ocean or sea that extends into the land; larger than a bay

19. **peninsula** – body of land nearly surrounded by water
20. **sea** – large bay of water partly or entirely surrounded by land
21. **dune** – mound, hill, or ridge of sand heaped by wind
22. **beach** – gently sloping shore of ocean or other water body
23. **river** – large stream of water that flows across land and empties into a larger water body
24. **mouth** – place where a river empties into a body of water
25. **delta** – land formed by deposits at the mouth of the river
26. **plateau** – area of high, flat land; larger than a mesa
27. **mesa** – high, flat landform rising steeply above surrounding land
28. **butte** – small, flat-topped hill; smaller than a mesa
29. **plain** – large area of level or gently rolling land
30. **valley** – V-shaped depression between mountains or hills
31. **channel** – narrow strip of water between two land bodies
32. **island** – body of land completely surrounded by water
33. **isthmus** – narrow strip of land (bordered by water) joining to larger bodies of land
34. **atoll** – ring-shaped coral island or string of islands
35. **lagoon** – shallow water area enclosed within an atoll or cut off from sea by strip of land

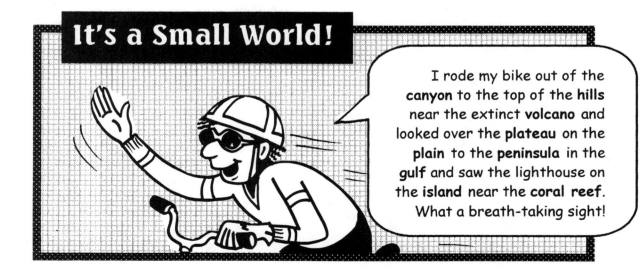

It's a Small World!

I rode my bike out of the **canyon** to the top of the **hills** near the extinct **volcano** and looked over the **plateau** on the **plain** to the **peninsula** in the **gulf** and saw the lighthouse on the **island** near the **coral reef**. What a breath-taking sight!

114

SURFACE FEATURES

Surface Changes

Conditions on Earth's surface are constantly changing the rocks and minerals that cover most of it. Some of the changes are slow; some happen very fast. All of them alter the composition, form, or location of materials on Earth.

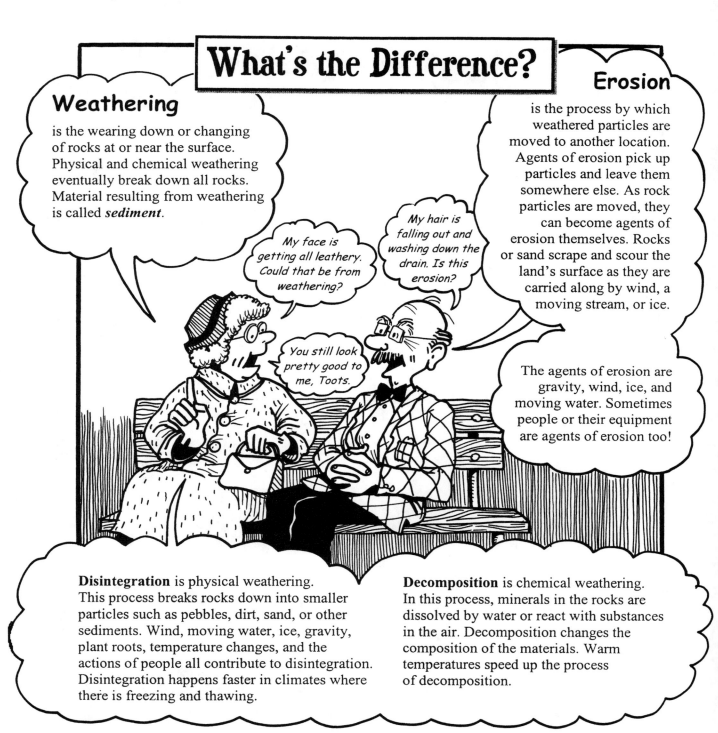

What's the Difference?

Weathering

is the wearing down or changing of rocks at or near the surface. Physical and chemical weathering eventually break down all rocks. Material resulting from weathering is called *sediment*.

My face is getting all leathery. Could that be from weathering?

My hair is falling out and washing down the drain. Is this erosion?

You still look pretty good to me, Toots.

Erosion

is the process by which weathered particles are moved to another location. Agents of erosion pick up particles and leave them somewhere else. As rock particles are moved, they can become agents of erosion themselves. Rocks or sand scrape and scour the land's surface as they are carried along by wind, a moving stream, or ice.

The agents of erosion are gravity, wind, ice, and moving water. Sometimes people or their equipment are agents of erosion too!

Disintegration is physical weathering. This process breaks rocks down into smaller particles such as pebbles, dirt, sand, or other sediments. Wind, moving water, ice, gravity, plant roots, temperature changes, and the actions of people all contribute to disintegration. Disintegration happens faster in climates where there is freezing and thawing.

Decomposition is chemical weathering. In this process, minerals in the rocks are dissolved by water or react with substances in the air. Decomposition changes the composition of the materials. Warm temperatures speed up the process of decomposition.

Gravity: An Agent of Erosion

Gravity is the force pulling everything downward. This pull causes movement of particles or loose material down a slope. The material that piles at the bottom of a slope is called *talus*.

Creep occurs when a mass of material moves very slowly downslope.

Slump occurs when layers of rock slip downslope leaving a curved scar.

A rockfall occurs when large masses of rock fall downslope quickly.

A landslide occurs when large amounts of material move quickly downslope.

A mudflow occurs when masses of debris and dirt mixed with rain slide downslope quickly.

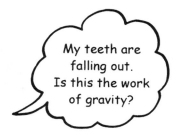

My teeth are falling out. Is this the work of gravity?

Wind: An Agent of Erosion

Wind (moving air) is a powerful agent of change on the face of the Earth. Its force picks up loose materials and moves them around, piling them in some places and whipping them around against rocks like scouring brushes against rocks. Wind is a major force for erosion in dry areas like deserts and areas where sand is plentiful, like shorelines.

Deflation is the process by which wind removes loose material from surfaces. In deflation, wind may pick up dirt, dust, sand, or topsoil. This is the cause of *dust storms* and *sandstorms*. Fine dust deposited by wind is called *loess*. Sand carried by the wind will be dropped when it comes up against an obstacle such as a bush or clump of grass. The result is a *sand dune*. When the wind sweeps all the material away in an area, the bare rock surface is called *desert pavement*. Sometimes the wind erodes soil and rock down to a level where water is present. Vegetation takes root and an *oasis* is formed.

Abrasion is the process in which particles in the wind scour and scrape against other surfaces. Rocks, cliffs, wood surfaces and other materials exposed to constant wind can become polished or pitted with this "sandblasting" action.

My nose is getting lumpy. Can I blame that on abrasion by the wind?

Dunes

A *sand dune* (drift or pile of sand) forms when the sand-carrying wind meets an obstacle. Hitting an obstacle causes the wind to slow, and the sand is dropped. Sometimes vegetation begins to grow on dunes and slows the sand's movement. If vegetation does not settle into a sand dune, the dune will continue to move with the wind. Dunes can build up as high as 100-1000 feet. They take different shapes and patterns, depending on the direction and intensity of the wind, the amount of vegetation, and the amount of sand.

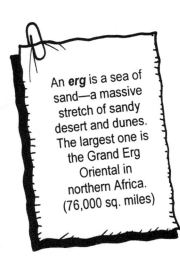

An **erg** is a sea of sand—a massive stretch of sandy desert and dunes. The largest one is the Grand Erg Oriental in northern Africa. (76,000 sq. miles)

Ice: An Agent of Erosion

Ice covers over 10% of Earth's surface permanently. Almost 75% of Earth's total supply of fresh water is frozen. The moving ice in glaciers and the ice that forms in cracks and crevices has great power to change Earth's surface.

Glaciers

A *glacier* is a large mass of ice in motion. A glacier forms when snow falls and doesn't melt. Over time, layers of snow accumulate into huge masses. The snow becomes compressed and heavy, and the snow is packed into clear, bluish *glacial ice*. The great weight causes slight melting at the bottom. The glacier slides and spreads along on this watery surface called *meltwater*. A glacier moves very slowly, but as it does, its weight and the debris it carries become powerful causes of erosion. The glacier slides along cracking, souring, and gouging the rock surface below.

Glacial flow is the movement of glacial ice.

A valley glacier (or alpine glacier) forms at high elevations.

A piedmont glacier is a valley glacier that spreads out onto a plain.

A continental glacier is a great mass of ice found near one of Earth's poles.

Icebergs are large blocks of ice that break from a glacier to float in the sea.

The Antarctic's Lambert Glacier is the world's longest glacier. (250 miles long)

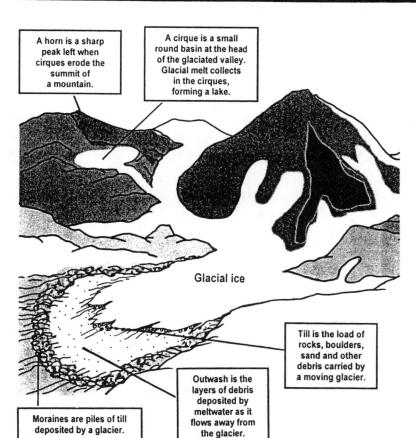

A horn is a sharp peak left when cirques erode the summit of a mountain.

A cirque is a small round basin at the head of the glaciated valley. Glacial melt collects in the cirques, forming a lake.

Glacial ice

Till is the load of rocks, boulders, sand and other debris carried by a moving glacier.

Outwash is the layers of debris deposited by meltwater as it flows away from the glacier.

Moraines are piles of till deposited by a glacier.

Glacial Erosion

Glaciers change the land as they move, scooping out formations such as cirques, horns, and hanging valleys, and leaving piles of debris.

Abrading is the scouring of rock surface that happens as a glacier moves over a surface carrying rough debris with it. Long scrapes, or striations, are left in the rock surface.

Plucking is the process of picking up fragments (large and small) and carrying them along.

Moving Water: An Agent of Erosion

Moving water is *the main agent of erosion*. Some of the most drastic and spectacular changes to Earth's surface are caused by runoff, rivers and streams, waterfalls, water underground, and the actions of ocean waves.

Runoff

Runoff is water from precipitation that flows across Earth's surface and eventually returns to streams, rivers, and the ocean. As it flows, this water moves soil and wears away rock. Runoff moves quickly on sloped land. On flat land, runoff tends to sink into the ground. Vegetation slows runoff because the roots of plants hold the water in the soil and keep soil from being carried away by the moving water. Areas with little soil, vegetation, or impermeable rock close to the surface particularly vulnerable to runoff.

1) In the water cycle, the heat from the sun evaporates some of the water from oceans, lakes, and rivers. Also, living things transpire water vapor into the air.
2) Air rises. Water vapor turns to liquid and forms clouds.
3) Water falls from the clouds as rain or snow.
4) Water soaks returns to the ground or into the bodies of water.

Rivers and Streams

Moving water has the power to carve through rock or carry amazing amounts of sediment and debris. Streams and rivers wear away rocks, move along dirt, rocks, and debris, and change the face of the Earth. The moving water does not do all the erosion alone. It is the load carried along by the water that does much of the work, scraping the underwater surfaces. When a river joins an ocean or lake, it slows down and drops much of the load it is carrying. This sediment builds up at the river's mouth and changes the land. When a river floods, the sediment it is carrying gets spread out from its banks, making more changes in the land. *(See pages 120-121 to learn more about the work of rivers and streams.)*

Groundwater

Some of the water that falls as precipitation slowly seeps down through soil into cracks in the rocks. This underground water, called *groundwater*, moves slowly beneath the surface through a series of interconnected pores in the rocks. Groundwater dissolves some of the rock and causes changes in Earth's lithosphere. It moves materials around, creating interesting formations and leaving deposits. *(See page 122 to learn more about the work of groundwater.)*

Oceans

Ocean waves carry the force of moving water. Often, their force is increased by the power of wind. Waves can move sand and rocks, or toss them against the shoreline. The force of waves constantly shapes and changes the coastline. When ocean water combines with fierce winds in a storm, great damage can be caused to coastal areas. *(See pages 125-128 to learn more about the work of the ocean.)*

Earth's Waters

Rivers & Streams

Most rivers begin as runoff that flows through ruts in Earth's surface. The water that does not soak into the ground flows in trickles that join other trickles, gradually forming larger streams of water. Eventually enough streams combine to form rivers. The streams or smaller rivers that join into a large one are called *tributaries*. Other rivers are formed from natural springs that come out of the ground through cracks in Earth's surface. Still others start from lakes or melting glaciers.

Get Sharp
Tip # 8
The beginning
of a river
is called
its *source*.

River Systems

As runoff grows into streams that merge into a river, the water flows in a pattern called a *drainage system*. The area drained by a river is the *drainage basin*. Each system is made up of a network of streams that form a pattern. Drainage patterns differ, depending on the kind of rock over which the river flows.

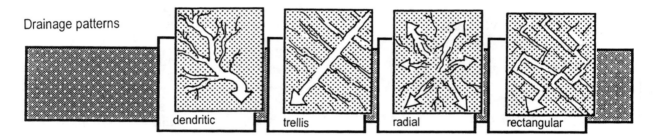

Drainage patterns

dendritic   trellis   radial   rectangular

Erosion by Rivers

The Nile River system is the longest in the world.

Rivers erode the land as they flow along. The force of the water wears rock away. Rocks, boulders, sand, and dirt can be pushed along by the water, scraping the river bottom as they move. The water also dissolves some materials in the riverbed. The amount of erosion caused by a river depends on the velocity of the river, the slope of the land, and the size of the load it carries. Slow-moving rivers carry less material than swift-moving rivers.

The riverbed is the path through which the river flows.

The river's load is the material carried by a river.

The suspended load is the debris picked up and carried along in the water.

The bed load is the material that is rolled along the river bottom because it is too heavy to be carried by the water.

Angel Falls in Venezuela is the highest waterfall in the world. It drops 3212 feet.

WOW!

Waterfalls

When water flows over an area of rock that has a layer of soft rock over a layer of hard rock, the soft rock will eventually wear away. If this results in the water flowing down a vertical rock surface, a *waterfall* is created.

River Deposits

Rivers don't hang onto all the sediment they carry. About 75% of the load a river has picked up is dropped before the river reaches the ocean. This sediment is dropped along the way in the beds of the rivers and streams that make up the river's drainage system.

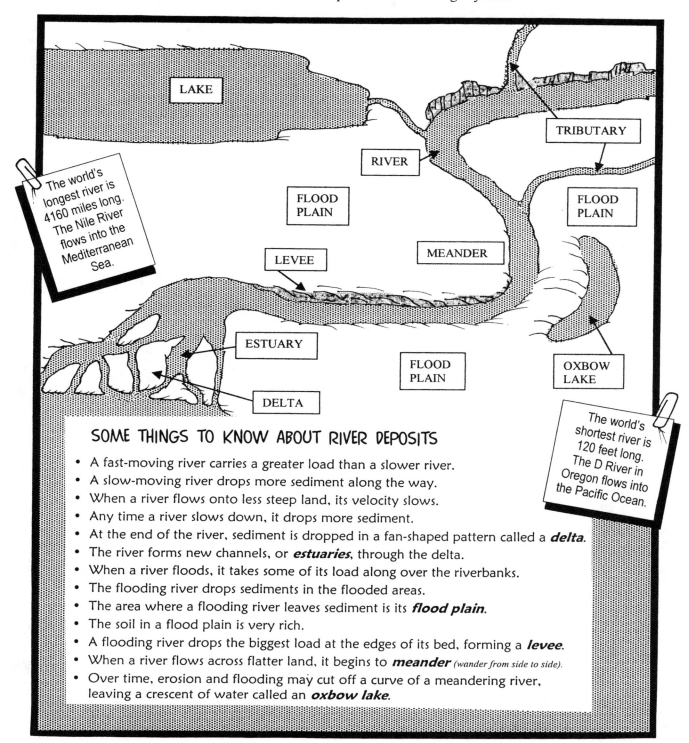

The world's longest river is 4160 miles long. The Nile River flows into the Mediterranean Sea.

The world's shortest river is 120 feet long. The D River in Oregon flows into the Pacific Ocean.

SOME THINGS TO KNOW ABOUT RIVER DEPOSITS

- A fast-moving river carries a greater load than a slower river.
- A slow-moving river drops more sediment along the way.
- When a river flows onto less steep land, its velocity slows.
- Any time a river slows down, it drops more sediment.
- At the end of the river, sediment is dropped in a fan-shaped pattern called a **delta**.
- The river forms new channels, or **estuaries**, through the delta.
- When a river floods, it takes some of its load along over the riverbanks.
- The flooding river drops sediments in the flooded areas.
- The area where a flooding river leaves sediment is its **flood plain**.
- The soil in a flood plain is very rich.
- A flooding river drops the biggest load at the edges of its bed, forming a **levee**.
- When a river flows across flatter land, it begins to **meander** *(wander from side to side)*.
- Over time, erosion and flooding may cut off a curve of a meandering river, leaving a crescent of water called an **oxbow lake**.

Better Grades & Higher Test Scores / SCIENCE
Copyright ©2003 by Incentive Publications, Inc., Nashville, TN.

Get Sharp: Earth's Waters

Groundwater

Some of the water on Earth's surface sinks into the ground, seeping into the soil and permeable rocks. This water becomes a part of a system of water beneath the ground, called **groundwater.** The action of groundwater causes many different deposits and natural formations.

Groundwater soaks down through soil and permeable rock until it reaches the layer of impermeable rock. The rocks just above the impermeable layers get soaked with water, forming a **zone of saturation**. The top edge of the zone of saturation is called the **water table**. Above the zone of saturation is the **zone of aeration**. In this area, the pores between rock particles are filled with air except during rainstorms. Permeable layers of rock that are filled with water are called **aquifers**. Gravity causes the movement of water through connecting pores in the aquifers.

Some groundwater returns to the surface in **springs**. The water flows out of the ground through a natural opening in the ground. Wells can be drilled to remove water from the water table or aquifers. For most wells, the water must be pumped to the surface. **Artesian wells** need no pump. In these cases, the groundwater flows because of pressure from the heavy weight of a column of water in the aquifer.

Get Smart Tip # 9
Permeable rocks have tiny spaces between particles. In **impermeable** rocks, particles are so close together that there are no spaces.

Hot Springs, Geysers, & Fumaroles, & Mudpots

Hot Springs are pools or streams of water that are heated naturally by heat from beneath Earth's surface. Water seeps down through rocks that are hot because they are near to pockets of hot magma. The hot water rises back to the surface and gathers in pools or running streams of hot water.

Fumaroles are vents in the ground through which volcanic gases and steam escape. Fumaroles usually appear in regions of former volcanic activity. **Solfataras** are fumaroles that give off sulfurous gases. **Mudpots** are areas where fumaroles bubble up through mud.

Geysers form in places where water seeps into the ground to an area where the surrounding rock is hot. The water gets heated and forms steam. The steam pushes the water up through cracks in the Earth with such force that it explodes in a shower of steam and water. The water settles back into the Earth until the steam builds up again.

Old Faithful in Yellowstone National Park is the world's best-known geyser. About every 76 minutes, it spouts gas and steam 120-150 feet in the air.

To learn more about groundwater, visit the kids' corner at www.groundwater.org

There are over 2 million cubic miles of fresh water beneath Earth's surface.

Erosion and Changes

Groundwater contains elements that can erode some of the softer rocks in and below the surface. The acid in the water dissolves away soft rocks like limestone, leaving the harder rock. This action creates some spectacular features.

Caves – When limestone is dissolved, some large openings are left in the harder rock. These are called caves or caverns. The process of cave-making takes thousands of years.

Natural Bridges – When some of a cavern roof collapses, but a part of the roof remains, a natural bridge is formed.

Sinkholes – Water drains through cracks in limestone, gradually dissolving it. This causes funnel-shaped depressions called sinkholes. The ground above the sinkhole can give way since there is no support left underneath it. This can cause serious damage to roads or buildings above the sinkhole.

Stalactites & Stalagmites – Water moving underground can contain large amounts of dissolved substances. When groundwater drips from the roof of a cave, it evaporates and leaves calcium carbonate deposits that look like giant icicles. Structures called *stalactites* develop and hang from cave ceilings. The structures that build up from the floor of the cave are *stalagmites*. Eventually, many stalactites and stalagmites join together into one column.

How to Make Your Own Stalactites and Stalagmites

I'm going to demonstrate how to make stalactites and stalagmites. Follow the directions carefully.

Voila!

You will need:
2 tall jars
a box of Epsom's salts
   or washing soda
a spoon
2 small rocks
cotton string
hot water
a flat plate

Did you know that the longest cave system in the world is in Kentucky? The Mammoth Caves are 352 miles long.

My hobby is spelunking, so I've seen lots of these in real caves.

1. Set the jars about a foot apart from each other.
2. Cut a piece of string long enough to reach between the jars and down into both jars.
3. Tie a small rock around each end of the string.
4. Pour hot water to fill each jar about half full.
5. Mix salt or soda into both jars. Add it until no more will dissolve.
6. Soak the string in one jar of water. String it from one jar to the other, with the ends in the jars.
7. Set the plate underneath the center of the string.
8. Make sure no one bumps or disturbs the materials.
9. Watch carefully to see what happens as the water evaporates.

Get Sharp: Earth's Waters

Lakes, Ponds, & Swamps

Lakes

A *lake* is an area of water completely surrounded by land. Most lakes contain fresh water, though some are filled with salt water. Lakes are usually formed from the flow of a river. In most cases, a lake will eventually disappear. It will be drained by rivers flowing out of it; or the water will evaporate; or the lake area will fill with vegetation and sediment.

Artificial Lakes

An artificial lake is one created by humans. Most artificial lakes are created when a dam is built across a river. The river water builds up behind the dam and floods the river valley. In most cases, a river is dammed for the purpose of producing electricity. The power of the flowing river is used to run generators and make electricity at whatever rate is needed.

Loch Ness is a lake in Scotland that is famous for something other than its size, depth, or water. It is well known because some people think a 50-foot long sea monster lives in the lake. This sea serpent is called the Loch Ness Monster.

Ponds

A *pond* is a small, shallow body of water. In most ponds, the sunlight can reach the bottom of the pond. This allows plant life to spread across the entire bottom surface.

Swamps

A *swamp* is a shallow body of water, usually poorly drained, found in a low-lying area. The water is filled with plant material that has fallen into the lake and decayed into thick, mucky layers of peat.

The Everglades is the largest swamp in the United States. It covers 5000 square miles.

The largest swamp in the world is in Brazil. The Grand Pantanal Swamp covers 42,000 square miles.

Oceans

Over three-fourths of the Earth's surface is covered with water (liquid or frozen). Most of this water (about 98% of it) is found in the world's oceans. An **ocean** is a large body of salt water. Sometimes oceans are called **seas**, although a sea is generally thought of as a body of salt water smaller than an ocean. There are five oceans: the Arctic, Atlantic, Pacific, Indian, and Southern (or Antarctic). The Pacific Ocean is the largest. The average ocean depth is two miles. The combined volume of all five oceans is 317 million cubic miles. All the oceans include smaller bodies of water around their edges, such seas, gulfs, and bays.

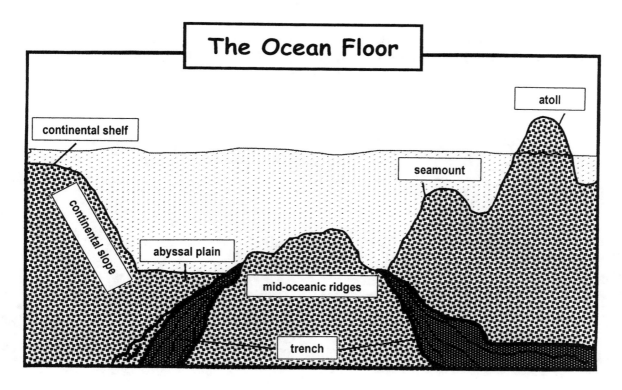

The bottom of the ocean is as varied as other parts of Earth's crust. It contains mountains, valleys, plains, crevasses, ridges, and volcanoes.

continental shelf - underwater land at the edges of the continents, from the shoreline to about 600 feet deep

continental slope - a steeper slope running downwards from the shelf to the ocean floor

abyssal plain - wide, flat area that makes up most of the ocean floor

mid-oceanic ridges - mountain ranges on the ocean floor

trench - a long, narrow crevasse in the ocean floor

seamount - a mountain with a peak below the water surface

atoll - a mountain on the ocean floor that breaks through the water surface

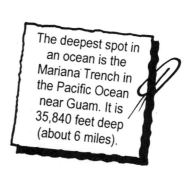

The deepest spot in an ocean is the Mariana Trench in the Pacific Ocean near Guam. It is 35,840 feet deep (about 6 miles).

Get Sharp: Earth's Waters

Water Movements

The water in the oceans is always moving, even if it looks still at the surface. Winds create waves (ripples of varying sizes) or currents. Earthquakes create giant waves called *tsunamis*. The gravitational pull of the Sun and Moon produce falling and rising movement called *tides*. Beneath the surface, circulation (vertical movement) of water is caused by wind patterns, Earth's rotation, and differences in temperature and salt content of the water.

Currents

Surface Currents are caused by winds blowing across the surface of the water. The wind whips water into motion, forming horizontal currents that flow in circular patterns. Wind currents move the upper levels of the ocean water, sometimes as far down as 600 feet or more. Earth's rotation adds to the circular pattern of the currents. The currents are interrupted or affected by the locations of the continents.

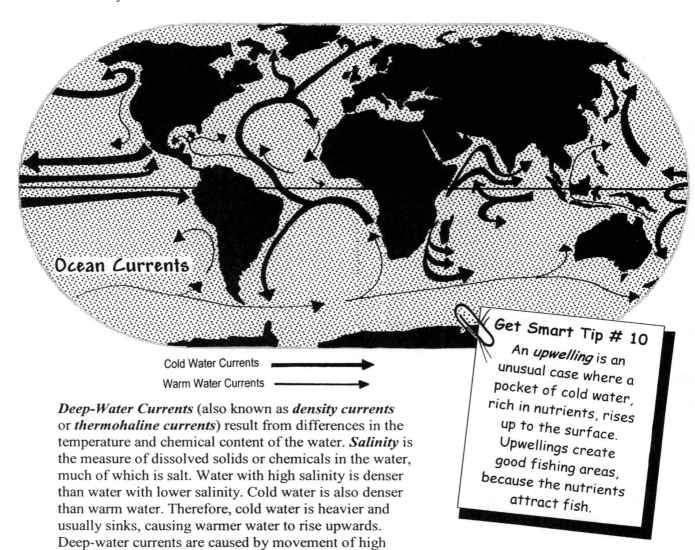

Ocean Currents

Cold Water Currents ⟶

Warm Water Currents ⟶

Deep-Water Currents (also known as *density currents* or *thermohaline currents*) result from differences in the temperature and chemical content of the water. *Salinity* is the measure of dissolved solids or chemicals in the water, much of which is salt. Water with high salinity is denser than water with lower salinity. Cold water is also denser than warm water. Therefore, cold water is heavier and usually sinks, causing warmer water to rise upwards. Deep-water currents are caused by movement of high density water into areas of lower density water.

Get Smart Tip # 10

An *upwelling* is an unusual case where a pocket of cold water, rich in nutrients, rises up to the surface. Upwellings create good fishing areas, because the nutrients attract fish.

Waves

Waves are ocean movements in which water rises and falls. Waves are caused by wind, earthquakes, or tides. Within a wave, water particles are moving in a circle. The height of a wave is equal to the diameter of the circle made by the particles moving in the wave.

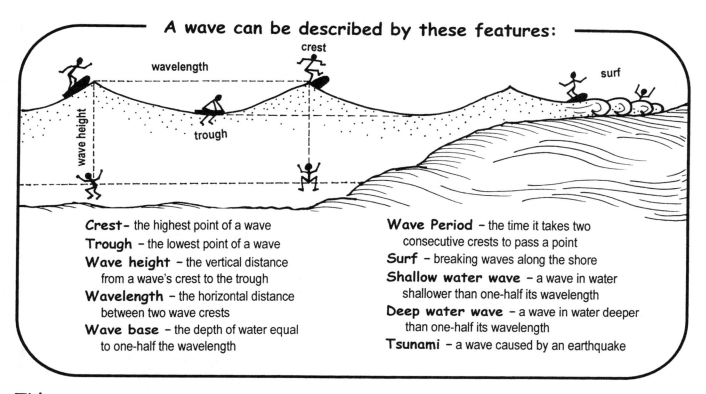

A wave can be described by these features:

Crest- the highest point of a wave

Trough – the lowest point of a wave

Wave height – the vertical distance from a wave's crest to the trough

Wavelength – the horizontal distance between two wave crests

Wave base – the depth of water equal to one-half the wavelength

Wave Period – the time it takes two consecutive crests to pass a point

Surf – breaking waves along the shore

Shallow water wave – a wave in water shallower than one-half its wavelength

Deep water wave – a wave in water deeper than one-half its wavelength

Tsunami – a wave caused by an earthquake

Tides

Tides are shallow water waves caused by the interaction among the gravity of Earth, Sun, and Moon. The Moon's gravitational force causes ocean water to bulge toward it, resulting in a *high tide*. Rotational forces cause another bulge on the opposite side of the Moon. The depressions between high tides are *low tides*. The *tidal range* is the difference between a high tide and a low tide. The relative positions of the Earth, Moon, and Sun affect the tides. During *spring tides*, high tides are at their highest and low tides at the lowest. *Neap tides* are minimum tides. The tidal range is greatest during spring tides, and least during neap tides.

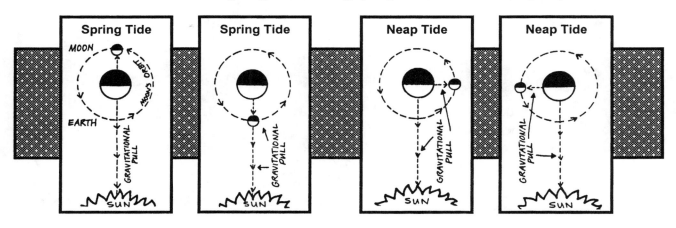

Better Grades & Higher Test Scores / SCIENCE
Copyright ©2003 by Incentive Publications, Inc., Nashville, TN.

Shorelines

A *shoreline* is the boundary where the land meets the water of the ocean. The *shore zone* is the area along the shore that lies between the point of high tide and low tide. There is constant movement in shore zones. Actions of waves move sand and other sediments, causing many changes along the ocean shore. *Longshore currents* (currents created when waves hit the beach at an angle) and *rip currents* (narrow currents that flow at a right angle to the shore) erode and change the shoreline.

Shoreline Features

Beaches are deposits of sand and other fragments left along the shoreline boundary. Beaches extend to 100 feet or more above and below the shoreline.

A sandbar builds up where longshore currents deposit sand and debris in deeper water parallel to the shore. The sandbar can be above water or covered by shallow water.

A spit forms along a curved shoreline where longshore currents drop loads of sand and debris extending across a bay.

A bay forms where part of the coastline is eroded.

A lagoon is a body of water cut off from the sea by a sandbar or reef.

Barrier islands develop from loads of sand and debris deposited parallel to the shore.

Arches and stacks are formations of resistant rock left standing after softer rock is worn away.

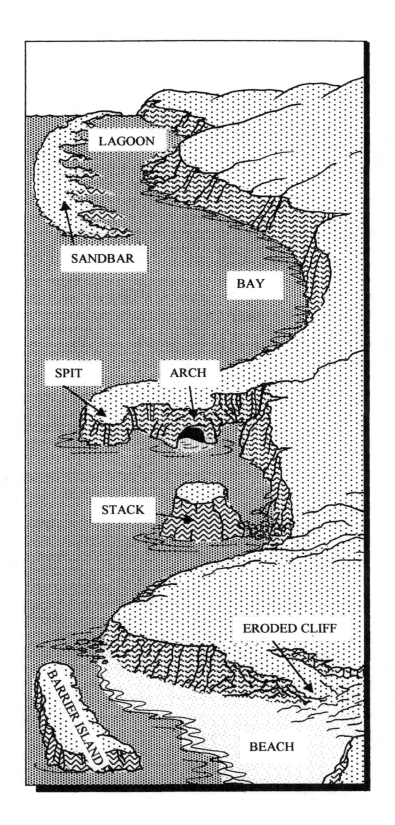

Earth's Atmosphere

The **atmosphere** is a blanket of air that surrounds the Earth. This envelope of air extends up to about 300 miles (480 kilometers) above Earth's surface; however, there is not a clear boundary where Earth's atmosphere ends. There are four layers in the atmosphere.

The troposphere (0-20 km) is the layer closest to Earth's surface. It contains most of the gases (including water vapor) and dust of the atmosphere. Clouds and weather occur in this layer. Jet stream winds occur at the very top boundary.

The tropopause (8-20 km) is the upper ceiling of the troposphere. It functions as a weather ceiling.

The stratosphere (20-50 km) is very cold, with temperatures of 0° to −50° C. The **ozone layer** is found in the stratosphere. Ozone is an important gas that absorbs most of the harmful ultraviolet radiation from the Sun.

The mesosphere (51-85 km) is the coldest layer of the atmosphere. Temperatures are as low as −100° C.

The thermosphere (above 85 km) is the uppermost layer. The air is thin, so energy from the Sun warms the air. The lower part, the **ionosphere**, has electrically charged particles useful for transmitting radio waves. The **exosphere** extends from about 500 km into outer space. The air is so thin here that some of its molecules escape into space.

Earth's Atmosphere

Altitude in Kilometers

- exosphere
- 500
- ionosphere
- thermosphere
- 90
- mesosphere
- 50
- stratosphere
- 20
- tropopause
- troposphere

Atmospheric pressure is the force of air pressing down on Earth's surface. At sea level, the atmospheric pressure is about 14.7 pounds per square inch. Air pressure varies in different places. A machine called a *barometer* is used to measure air pressure. Air pressure is affected by temperature and altitude.

What is air, anyway?

It is a mixture of gases. The air is 78% nitrogen, 21% oxygen, and 1% other gases.

129

Winds

Wind is moving air. Air always moves from areas of high atmospheric pressure to areas of low pressure. Air moves in great spirals from high to low pressure, creating giant wind systems around the Earth. Air also moves horizontally across Earth's surfaces. The differences in surface temperatures can also affect air movement.

Winds and Wind Systems

polar easterlies – These dry, cold air currents move from northeast to southwest over the higher latitudes (60° to 90°) of the Northern Hemisphere and from southeast to northwest over the polar zones of the Southern Hemisphere (60° -70° south latitude).

westerlies – In the middle latitudes (30° to 60° north or south of the equator), the prevailing winds blow from west to east.

trade winds – These winds blow toward the equator from a latitude of about 30° north and south of the equator. They blow from northeast to southwest in the Northern Hemisphere and from southeast to northwest in the Southern Hemisphere.

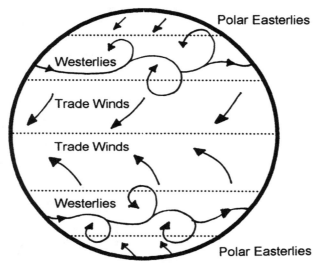

Get Sharp Tip # 11
The rotation of Earth causes air to move fastest at the equator. This creation of eastward winds just above and below the equator is called the *Coriolis Effect.*

doldrums – At the equator, there is little horizontal movement of air. In this windless zone, the air is moving upwards, leaving a seemingly windless area.

sea breeze – During the day, land absorbs heat from the Sun. The warm air rises and cool sea air blows onto the land.

land breeze – At night, the land loses its heat, and the warmer air over the water rises. Cooler land air moves out to sea to fill in the space.

jet streams – These narrow belts of fast-moving air flow in a westerly direction at the higher levels of the troposphere.

An instrument called an **anemometer** measures the speed, or velocity, of the wind. Velocity is measured in miles or kilometers per hour. A scale for measuring wind force was developed in 1806 by Sir Francis Beaufort. This is the scale, with an added description of what action might be seen for each level of wind.

Some Local Winds

Monsoons - stormy winds that bring heavy rain to southern Asia and northern Australia

The rainy season, which lasts from about June to September, is called the Monsoon season.

Chinooks - warm, dry winds that flow down the side of a mountain range.

These are called "foehns" in Europe.

Siroccos - winds that blow north from the Sahara Desert

The winds pick up moisture over the Mediterranean Sea and bring warm rain to Europe.

Harmattans - cool dry winds that blow west from Sahara Desert

They bring relief from the intense heat to the tropics of Northern Africa.

Typhoons - the name given to hurricanes in the northwest Pacific Ocean

Cyclones - the name given to hurricanes in the Pacific Ocean north of Australia and in the Indian Ocean

The Beaufort Scale of Wind Strength

Beaufort Number	Wind Name	Speed mph	Description
1	light air	1-3	wind direction shown by smoke drift; weather vanes inactive
2	light breeze	4-7	wind felt on face; leaves move slightly; weather vanes active; smoke does not rise vertically
3	gentle breeze	8-12	leaves and small twigs move constantly; flags blow
4	moderate breeze	13-18	dust and paper blow about; twigs and thin branches move
5	fresh breeze	19-24	small trees sway; white caps form on lakes
6	strong breeze	25-31	large branches move; telegraph wires whistle; umbrellas are hard to control
7	moderate gale	32-38	large trees sway; it's somewhat difficult to walk
8	fresh gale	39-46	twigs break off of trees; walking against the wind is very difficult; possible damage to property
9	strong gale	47-54	slight damage to buildings; shingles are blown off roof
10	whole gale	55-63	trees are uprooted; much damage to buildings
11	storm	64-75	widespread damage
12	hurricane	over 75	extreme destruction

Hurricanes are given men's or women's names. There are 6 sets of names for hurricanes. The lists rotate every 6 years. *The names on the list for 2002 are Ana, Bill, Claudette, Danny, Erika, Fabian, Grace, Henri, Isabel, Juan, Kate, Larry, Mindy, Nicholas, Odette, Peter, Rose, Sam, Teresa, Victor, and Wanda.*

Weather

Weather is the state of the atmosphere at any one time. The factors involved in weather are temperature, air pressure, wind, and moisture.

Humidity is the ability of air to hold moisture. Air with high humidity has a high moisture content.

Relative humidity is a measurement of the amount of water vapor in the air in relationship to the total volume of the air.

Dew point is the temperature at which water vapor condenses in a layer of air.

Some "Air-y" Facts

- Cold air sinks because it is denser than warm air.
- Dense air creates areas of high pressure.
- Warm air is less dense than cold air.
- Areas of low pressure are created when air warms.
- Warm air can hold more moisture than cold air.
- When air cools, it has to drop some of its moisture.
- Warm air rises, losing moisture as it cools.
- Cold air sinks and pushes warm air up.
- When the relative humidity of an amount of air reaches 100%, water vapor condenses into water.

Get Smart Tip # 12

When air masses meet, expect stormy weather at the **front**!

Clouds & Precipitation

Clouds are groups of billions of tiny droplets of water in the air. They are formed when warm air rises and cools, or when warm air meets cold air. Since cooler air holds less water vapor than warm air, some of the vapor condenses. When the cloud becomes too full of water drops, water will fall from the cloud in different forms of *precipitation* (such as rain or snow).

Air Masses & Weather Fronts

When air slows over an area of land or water, a large body of air called an *air mass* may form. Air masses differ in temperature, atmospheric pressure, moisture content, and pattern of air circulation. In low-pressure air masses (called *cyclones*), air moves counterclockwise toward the center. In masses of high pressure (called *anticyclones*), air circulates clockwise out from the center.

When air masses meet, they form a boundary that keeps them separate. This boundary is called a *front*. Different kinds of fronts can form, depending on the nature of the air masses that are meeting.

> *Each different front that develops creates a particular kind of weather pattern.*

A cold front develops when a cold air mass invades a warm air mass. The heavier cold air sinks and slides under the warm air, pushing it steeply upwards. This causes cumulus and cumulonimbus clouds to form. Rainstorms or thunderstorms develop.

A warm front develops when a warm air mass invades a cold air mass. The less dense warm air slides up and over the cold air. Cirrus clouds, altostratus clouds, and nimbostratus clouds develop. Rain or snow often accompanies warm fronts.

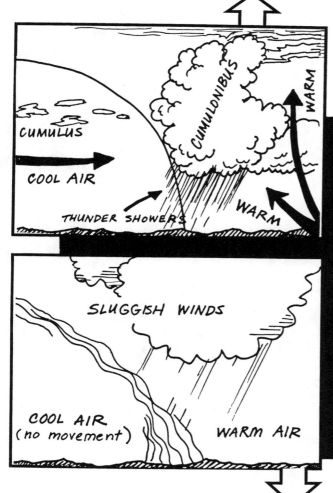

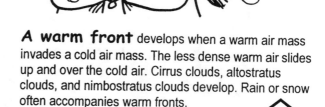

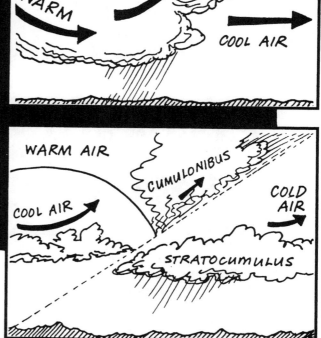

A stationary front develops when a cold front or a warm front stays in place for several days without invading another front. Clouds and precipitation often form at the boundary.

An occluded front develops when two masses of cold air meet. The cold air forces warmer air caught between the two fronts upwards. Cumulonimbus and stratocumulus clouds usually form. High winds and heavy rain or snow may result.

133

TYPES OF CLOUDS

Cumulus	turret-shaped tops, flat bottoms	fair weather
Cumulonimbus	thunderheads (large, dark cumulus)	thunderstorm
Stratus	smooth layers of low clouds	chance of drizzle or snow
Stratocumulus	piles of clouds in layers	chance of drizzle or snow
Nimbostratus	smooth layers of dark gray clouds	continuous precipitation
Altostratus	thick sheets of gray or blue clouds	rain or snow
Altocumulus	piles of clouds in waves	rain or snow
Cirrus	feather-like clouds (made of ice crystals)	fair weather
Cirrostratus	thin sheets of clouds (causes halo around the Sun or Moon)	rain or snow within 24 hours
Cirrocumulus	"cottony" clouds in waves	fair weather

Weather Conditions

Expect the **thunderstorm** to show itself with lightning and thunder, and possibly heavy rains.

This weekend, water will freeze on ice pellets in the clouds and make crystals, which will join with other crystals and produce **snow**.

Water droplets are forming and hovering over the ground causing **fog**. This has formed because the air above ground is being cooled by ground with a cooler temperature.

The surface temperature is below freezing this morning, so the water vapor in the air will freeze as it touches the ground and other surfaces, leaving **frost**.

Frozen raindrops keep blowing upwards in air currents. Water keeps freezing around each icy stone until it is ready to fall to earth as **hail**.

You can expect **dew**. The ground will get cold enough for water vapor in the air to condense into water droplets. These will form on surfaces such as leaves or ground.

Wild winds and heavy, driving snow have combined to cause the **blizzard** we're experiencing.

Whirling funnels or air, called **tornadoes**, are expected to form between the bottom of a storm cloud and the ground.

Tiny droplets of water vapor in the clouds will join together to make bigger drops today. Before long, these drops will be falling to the ground as **rain**.

A **hurricane** has developed over the warm, tropical ocean and is hitting with wind strength over 75 mph.

Right now raindrops are falling through a layer of air colder than 3° C. This is freezing rain, or **sleet**.

There has been no rain or any form of precipitation for weeks. We are officially in a **drought** period.

Better Grades & Higher Test Scores / SCIENCE
Copyright ©2003 by Incentive Publications, Inc., Nashville, TN.

Get Sharp: Weather

Climate

Climate is the average long-range weather of an area. Average temperatures, amounts and kinds of precipitation, and wind patterns are all a part of the climate of an area. Different parts of the world certainly have different climates. There are many factors that work together to form the climate conditions of a particular place. Latitude, topography, winds, ocean currents, altitude, masses of land, and bodies of water all affect climate.

Earth's Revolution and Tilt

The Earth's orbit is elliptical. Because of this, Earth is closer to the Sun during parts of its orbit than at other times. Earth's revolution, in combination with Earth's tilt, causes the seasons and the climate changes that come with the seasons.

Summer Solstice – On this day, one of Earth's poles is tilted most directly toward the Sun. This occurs in the Northern Hemisphere on June 21 or 22. Areas north of the Arctic Circle have 24 hours of daylight. Summer solstice occurs in the Southern Hemisphere on December 21 or 22. Areas south of the Antarctic Circle have 24 hours of daylight.

Fall Equinox – On this day, Earth's tilt is sideways toward the Sun, so the hours of daylight and darkness are the same in both hemispheres. In the Northern Hemisphere, this occurs on September 22 or 23. In the Southern Hemisphere, this occurs on March 20 or 21.

Winter Solstice – On this day, one of Earth's poles is tilted most directly away from the Sun. In the Northern Hemisphere, this occurs on December 21 or 22. Areas north of the Arctic Circle have 24 hours of darkness. In the Southern Hemisphere, this occurs on June 21 or 22. Areas south of the Antarctic Circle have 24 hours of daylight.

Spring Equinox - On this day, Earth's tilt again is sideways toward the Sun, so the hours of daylight and darkness are the same in both hemispheres. In the Northern Hemisphere, this occurs on March 20 or 21. In the Southern Hemisphere, this occurs on September 22 or 23.

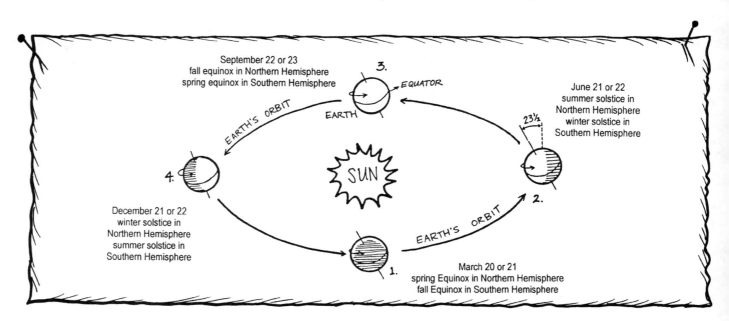

September 22 or 23
fall equinox in Northern Hemisphere
spring equinox in Southern Hemisphere

3.

EQUATOR

EARTH'S ORBIT

EARTH

June 21 or 22
summer solstice in
Northern Hemisphere
winter solstice in
Southern Hemisphere

23½

SUN

4.

December 21 or 22
winter solstice in
Northern Hemisphere
summer solstice in
Southern Hemisphere

EARTH'S ORBIT

2.

1.

March 20 or 21
spring Equinox in Northern Hemisphere
fall Equinox in Southern Hemisphere

GET SHARP →

on Life Science

Characteristics of Life

What is Life?

All living things, or **organisms**, have some things in common.
Every living thing has all these characteristics.

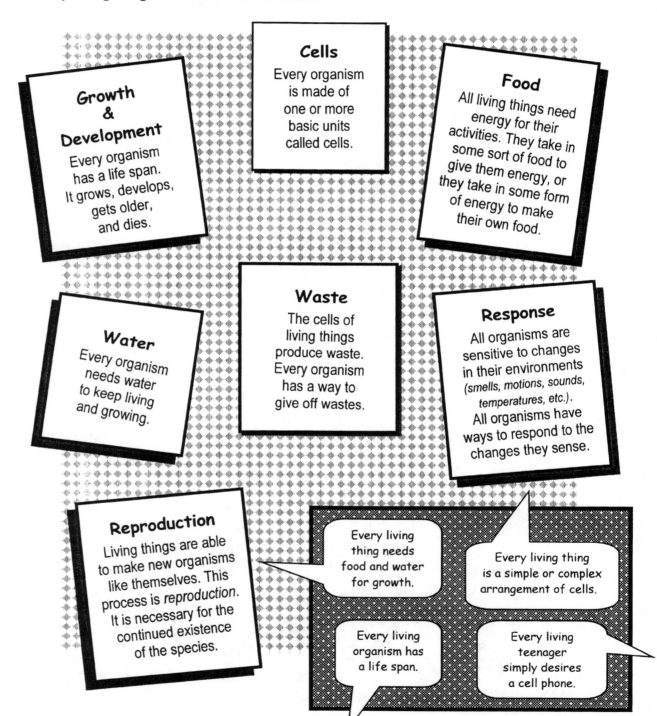

Growth & Development
Every organism has a life span. It grows, develops, gets older, and dies.

Cells
Every organism is made of one or more basic units called cells.

Food
All living things need energy for their activities. They take in some sort of food to give them energy, or they take in some form of energy to make their own food.

Water
Every organism needs water to keep living and growing.

Waste
The cells of living things produce waste. Every organism has a way to give off wastes.

Response
All organisms are sensitive to changes in their environments (*smells, motions, sounds, temperatures, etc.*). All organisms have ways to respond to the changes they sense.

Reproduction
Living things are able to make new organisms like themselves. This process is *reproduction*. It is necessary for the continued existence of the species.

Every living thing needs food and water for growth.

Every living thing is a simple or complex arrangement of cells.

Every living organism has a life span.

Every living teenager simply desires a cell phone.

Cells are the basic units of every living organism. There are many different kinds of cells, but they all share some basic features. Plant cells and animal cells differ somewhat. Plant cells have some features that animal cells do not have (vacuoles, cell walls, chloroplasts). Plant cells are usually larger and more oblong than animal cells.

Animal Cell

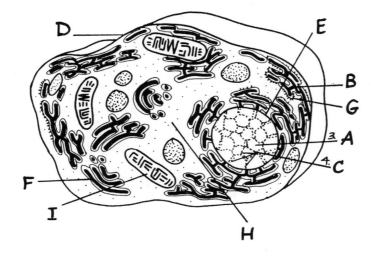

A. nucleus
the brains of the cell; carries the codes that control all cell activities

B. ER (endoplasmic reticulum)
a system of tubes where many cell substances are made

C. chromosomes
carry the code material (DNA) that directs cell activities

D. cell membrane
controls movement of materials in and out of the cell

E. nuclear membrane
controls movement of materials in and out of the nucleus

F. Golgi bodies
assemble, release, and store some chemicals

G. ribosomes
manufacture proteins

H. cytoplasm
substance that holds all other parts in suspension

I. mitochondria
release energy for food

J. lyosomes
gobble up waste materials

K. vacuole
stores water

L. cell wall
gives shape and support to plant cells made of strong substance (cellulose)

M. chloroplasts
contain chlorophyll which traps sunlight to help the plant make food

Plant Cell

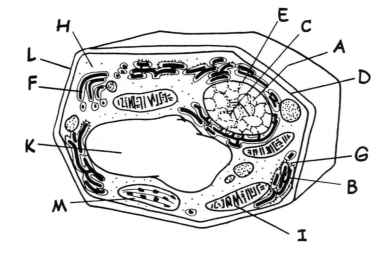

Life Processes

Plasmolysis and osmosis may sound like weird diseases. They're not! They are two of the many processes that take place in living cells. Here's a brief glossary that will help you sort out the different cell processes.

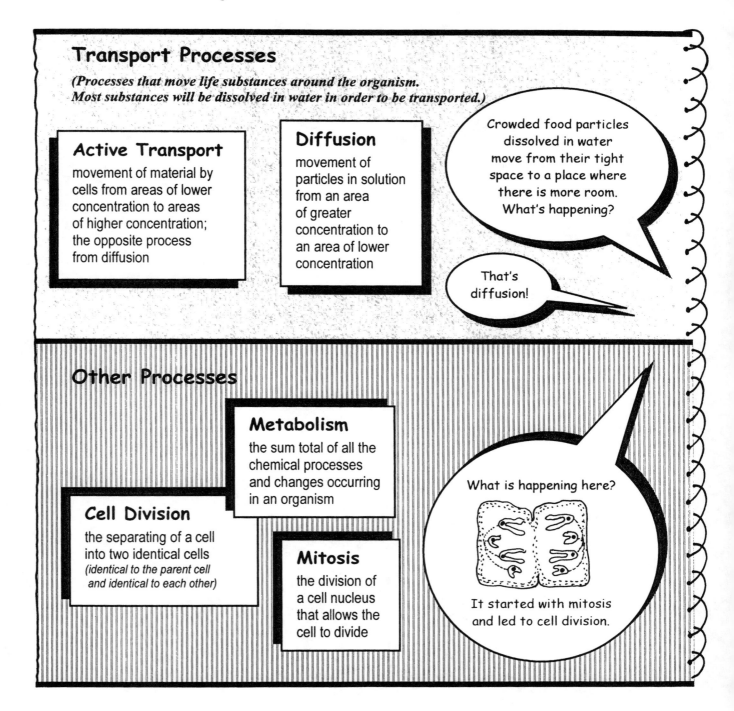

Transport Processes

(Processes that move life substances around the organism. Most substances will be dissolved in water in order to be transported.)

Active Transport

movement of material by cells from areas of lower concentration to areas of higher concentration; the opposite process from diffusion

Diffusion

movement of particles in solution from an area of greater concentration to an area of lower concentration

Crowded food particles dissolved in water move from their tight space to a place where there is more room. What's happening?

That's diffusion!

Other Processes

Metabolism

the sum total of all the chemical processes and changes occurring in an organism

Cell Division

the separating of a cell into two identical cells *(identical to the parent cell and identical to each other)*

Mitosis

the division of a cell nucleus that allows the cell to divide

What is happening here?

It started with mitosis and led to cell division.

That's life!

Transport Processes

Cells in a plant root take in water and nutrients through the cell membranes. What's going on?

That's osmosis!

Osmosis
diffusion of water through a membrane

A plant goes limp when water diffuses out of the cells. What's causing this?

That's plasmolysis!

Plasmolysis
shrinking of the cytoplasm in cells due to loss of water

Other Processes

A swimmer sits on a raft in the wind. Her body shivers to warm up her muscles. What's going on here?

That's homeostasis!

Homeostasis
the tendency of an organism toward balance; a process of adjusting other processes in the cells so the organism is in balance

Respiration
the process in which cells release energy from food

What's going on here?

It looks like reproduction to me!

Reproduction
the process in which organisms create other organisms identical to themselves

— 141 —

Life Classification

In order to study living organisms, scientists divide them into groups with characteristics that are similar. One system of classification divides organisms into five groups called kingdoms. The kingdoms are then subdivided into smaller groups (*phylum, class, order, family, genus, species*). The largest groups in the plant kingdom are called *divisions*. The smallest category for classification is a *species*. Organisms in a species usually reproduce only with other members of the same species. The scientific name of a species uses the Latin name of the genus and the species. For example, *Canis familiaris* is the scientific name for a domestic dog.

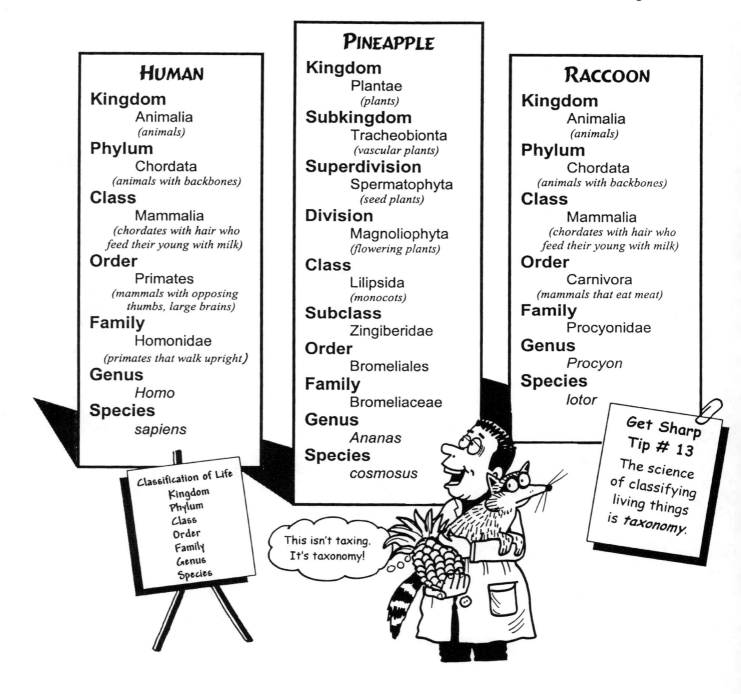

HUMAN

Kingdom
Animalia
(animals)
Phylum
Chordata
(animals with backbones)
Class
Mammalia
(chordates with hair who feed their young with milk)
Order
Primates
(mammals with opposing thumbs, large brains)
Family
Homonidae
(primates that walk upright)
Genus
Homo
Species
sapiens

PINEAPPLE

Kingdom
Plantae
(plants)
Subkingdom
Tracheobionta
(vascular plants)
Superdivision
Spermatophyta
(seed plants)
Division
Magnoliophyta
(flowering plants)
Class
Lilipsida
(monocots)
Subclass
Zingiberidae
Order
Bromeliales
Family
Bromeliaceae
Genus
Ananas
Species
cosmosus

RACCOON

Kingdom
Animalia
(animals)
Phylum
Chordata
(animals with backbones)
Class
Mammalia
(chordates with hair who feed their young with milk)
Order
Carnivora
(mammals that eat meat)
Family
Procyonidae
Genus
Procyon
Species
lotor

Classification of Life
Kingdom
Phylum
Class
Order
Family
Genus
Species

This isn't taxing. It's taxonomy!

Get Sharp Tip # 13
The science of classifying living things is *taxonomy*.

The Five Kingdoms

There are more than 2 million different kinds of living things on Earth.

monera

The moneran kingdom consists of two phyla, both made up of one-celled organisms. **Bacteria** mostly absorb food. Some contain chlorophyll. Bacteria can be round, rod-shaped, or spiral in shape. **Cyanobacteria** make their own food, contain chlorophyll, and are mostly blue-green in color.

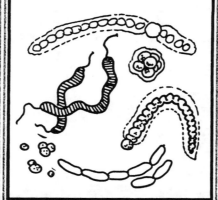

protista

The protist kingdom consists of several phyla including *ciliates, euglenoids, golden algae, sporozoans, sarcodines, slime molds,* and *flagellates.* Protists are mostly one-celled organisms. Some make their own food, but most take in or absorb food. Most protists move with the help of flagella, pseudopods, or cilia.

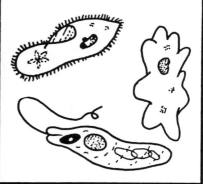

animalia

The animal kingdom consists of several phyla. Animals are many-celled organisms that must find food because they cannot make their own. Most animals are able to move from place to place at some point in their life span. *(See pages 152-153 for more on animal classification.)*

A *virus* is a complex organic compound that has some characteristics of both living and nonliving things. A virus is not made of cells, and can survive only inside living cells of other organisms.

fungi

The fungus kingdom consists of three phyla. All fungi absorb food. **Sac fungi** can be one or many cells. Spores are produced in a small sac called an ascus. **Club fungi** have many cells and produce spores in club-shaped sacs called basidia. **Sporangium fungi** have many cells and produce spores in sporangia.

plantae

The plant kingdom consists of several divisions. Plants are many-celled organisms containing a material called cellulose. Most plants are able to make their own food. Plants are generally anchored in place and do not move from place to place. *(See pages 144-145 for more on plant classification.)*

Better Grades & Higher Test Scores / SCIENCE
Copyright ©2003 by Incentive Publications, Inc., Nashville, TN.

Get Sharp: Life Classification

Plant Classification

The plant kingdom has over 300,000 different species—
from simple algae to complex trees and flowering plants.
Plants are classified into major divisions. In four of the divisions,
the plants are **nonvascular plants**. These are plants without vessels.
Most of them absorb water without the help of roots, stems, or leaves.

The tallest and longest-living organisms on Earth are plants. The tallest plant is a coast redwood tree standing 363 feet. A bristle-cone pine tree can live to be 5000 years old!

Nonvascular Plants

CHLOROPHYTA
(Green Algae)

This division contains one-celled plants living in colonies, or many-celled, green plants living in water or on land. Green algae make their own food. They reproduce by forming zygospores.

PHAEOPHYTA
(Brown Algae)

This division contains many-celled brown plants. They make their own food and live mostly in salt water.

RHODOPHYTA
(Red Algae)

This division contains many-celled red plants. They make their own food and live mostly in deep salt water.

BRYOPHYTA
(Mosses & Liverworts)

This division contains many-celled green plants. They make their own food and have a root system of **rhizoids**. They grow in moist areas on land and reproduce from spores in capsules.

Here, in Hawaii, we eat 40 different species of algae.

We'll have the seaweed soup, please.

144

Vascular plants are plants with vessels. The vessels in plants are a system of tube-like structures that move water with nutrients to all parts of the plant.

Vascular Plants

TRACHEOPHYTA

CLUB MOSSES & HORSETAILS

Club mosses and horsetails are simple vascular plants. They have many cells and live on land. They are green and make their own food. Both reproduce by forming spores in sporangia.

Most plants make their own food, and don't need to find food anyplace else. But Venus's fly-trap is a flesh-eating plant. It cleverly traps insects for its dinner.

FERNS

Ferns are many-celled vascular plants. They are green and make their own food. Ferns have feathery leaves called fronds. They live on land or in water. They reproduce by forming spores in sporangia.

SEED PLANTS

Seed plants are complex vascular plants with roots, stems, leaves, and seeds. They reproduce by means of seeds that are produced inside a fruit or in cones.

Monocots have a single cotyledon (seed leaf) inside their seeds. The leaves are narrow with parallel veins. The flower petals and sepals grow in multiples of 3.

ANGIOSPERMS

Angiosperms are plants that produce seeds inside a fruit. They are also called *flowering plants*. Angiosperms have two classes: monocots and dicots.

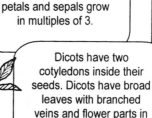

Dicots have two cotyledons inside their seeds. Dicots have broad leaves with branched veins and flower parts in multiples of 4 or 5.

GYMNOSPERMS

Gymnosperms are plants that produce seeds outside of a fruit. Most produce seeds inside cones. (These plants are called *conifers*.)

145

Plant Processes

Plants don't just stand around. They're very busy most of the time doing important work, such as growing, "breathing," and making food. In the process of doing their work, they provide some important supplies for the other living things on the planet.

Photosynthesis

The chlorophyll in the green plant cells captures sunlight. The plant uses the energy of the light to combine carbon dioxide with water, producing sugar and oxygen. The sugar is the food the plant will use for its growth and other processes. After it is made, the sugar is stored until it is needed. The oxygen produced is released into the air.

Gas Exchange

Tiny openings in the leaf (stomata) allow gases to pass in and out of the leaf by diffusion. Guard cells surrounding each stoma control them, opening them during the day and closing them at night. In gas exchange, carbon dioxide, oxygen, and water vapor pass in and out of the leaves, as the plant needs to use or release them.

Respiration

Respiration is the opposite of photosynthesis. The plant uses oxygen to break down sugar, releasing energy for its life processes. Water and carbon dioxide are produced as respiration takes place.

Transpiration

Plants lose water vapor through the stomata in the leaves. A plant takes in water through its roots and loses a considerable amount each day through transpiration.

Plant Behaviors

What's the Difference?

I get stimulus and response confused. What's the difference?

A **stimulus** is something happening in the environment that affects the behavior of an organism.

A **response** is a change in the organism's behavior as a result of the stimulus. Plant responses are called **tropisms**.

Do you mean plants can behave?

Yes, plants behave in response to things that happen around them.

What things?

A stimulus is something such as: light, absence of light, touch, temperature, changes in air quality, water, and gravity.

How do plants respond?

A plant might close up when touched. A flower might wither due to cold temperatures. Roots grow downwards toward gravity. This is called **geotropism**.

A stem can grow towards the light. This is **phototropism**. A flower might bloom as the days grow longer. This is **photoperiodism**.

There are **negative tropisms**, too. For example, roots grow the opposite direction from light.

Get Sharp: Plant Behaviors

Plant Structures

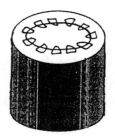

PLANT TISSUES

xylem is made up of vessels that move water and nutrients around the plant.

phloem has cells like tubes. They transport food from the leaves to other parts of the plant.

cambium is a tissue that makes plant stems grow thicker. It makes new phloem and xylem cells.

Most plant species have three main organs: roots, stems, and leaves. Each organ serves particular functions for the plant.

Roots

Roots are the plant's anchors. They hold the plant firmly in the ground so it can't be moved by wind or water. The cells and tissues in roots absorb water and nutrients from the soil.

Taproots – A taproot system is found in plants such as carrots, radishes, beets, or dandelions. These plants have a long, thick, main root that stores food.

Fibrous Roots – Fibrous root systems are made up of branches getting ever smaller as they reach out beneath the plant. Fibrous roots, such as in grass and trees, keep soil in place.

Stems

The stem is a plant's support system, holding the leaves upright. Tissues in the stems also transport food and water around the plant.

Herbaceous stems are green and soft. Most plants with herbaceous stems are *annuals* (plants that grow, reproduce, and die all in one year). Beans, petunias, and zucchini plants have herbaceous stems.

herbaceous stem

woody stem

Woody stems are rigid and hard, with a covering like bark. Most plants with woody stems are *perennials* (plants that do not die after one growing season). Trees and lilac bushes have woody stems.

Leaves

Leaves are the plant's food factories. Leaves trap the sunlight for use in photosynthesis to make food. Other important plant processes, such as respiration, transpiration, and gas exchange, also take place in the leaves. (See page 149.)

The blade of the leaf is the part the traps the sunlight.

The petiole is the leaf stalk that attaches the leaf to the plant stem.

The epidermis of a leaf is the thin layer of cells that covers the outer surfaces of the leaf, protecting the inner leaf parts.

Stomata are tiny openings in the epidermis that allow air and water to move in and out of the leaf. Each stoma is controlled by *guard cells* that surround the opening and cause it to open and close.

148

Plant Reproduction

Reproduction is part of the life cycle of every plant. It is the way plants create new organisms identical to themselves. The methods of plant reproduction differ, depending upon the simplicity or complexity of the plant. Some methods of reproduction are *sexual* (involving material from two different parent cells). Others are *asexual* (involving only one parent cell).

Reproduction Without Seeds

Asexual reproduction is found in some of the simplest plants. Only one parent produces an offspring. In some algae, pieces of a parent plant break off and grow into new individual plants. In other algae, reproduction occurs when a cell from the parent plant divides, forming beginning cells for a new plant. In some plants, new offspring can sprout up from underground roots called *rhizomes* or *runners*.

Conjugation is a method of reproduction that results in a zygospore. In some simple plants, cytoplasm from the cell of one plant moves into a cell of another plant. Material from the nuclei of two cells combine (fuse) and form a new cell called a zygospore. This zygospore eventually develops into a new plant.

Now, that's interesting!

Reproduction with spores occurs in many seedless plants such as mosses, horsetails, and ferns. A parent plant produces thousands of spores from sex cells called *gametes*. In ferns, when female cells (eggs) combine with male cells from spores (sperm), a *zygote* is formed. It divides and grows into a new fern plant.

Rhizomes are roots that grow underground from some plants. Sometimes new plants spring up from rhizomes.

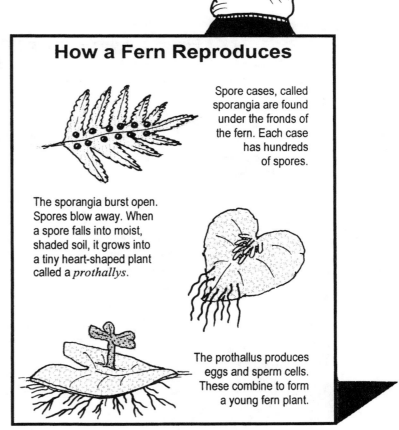

How a Fern Reproduces

Spore cases, called sporangia are found under the fronds of the fern. Each case has hundreds of spores.

The sporangia burst open. Spores blow away. When a spore falls into moist, shaded soil, it grows into a tiny heart-shaped plant called a *prothallys*.

The prothallus produces eggs and sperm cells. These combine to form a young fern plant.

Reproduction in Seed Plants

Many plants reproduce through *seeds*. A seed is a fertilized plant egg. Seed plants have female and male reproductive organs. These organs are found in flowers (in angiosperms) or in cones (in most gymnosperms).

Here's how reproduction works in seed plants:

- **Eggs** form inside female plant cells called *ovules*.

- **Sperm** form in male cells called *pollen grains*.

- **Pollination** occurs when a pollen grain transfers to an ovule.

- **Fertilization** occurs after pollination when a sperm cell joins with one of the eggs inside the ovule.

- **A zygote** is the fertilized egg that results from fertilization.

- **A seed** forms from the fertilized ovule.

- **An embryo** (young plant) begins to grow within the seed. The hard covering of the seed protects the young plant as it grows.

- **Dormancy** is a resting period that occurs for many seeds after fertilization. During the dormant period, the embryo does not grow.

- **Germination** is the process where the seed begins to grow from an embryo into a young plant called a *seedling*. Germination occurs only when temperature, moisture, oxygen, and soil conditions are just right.

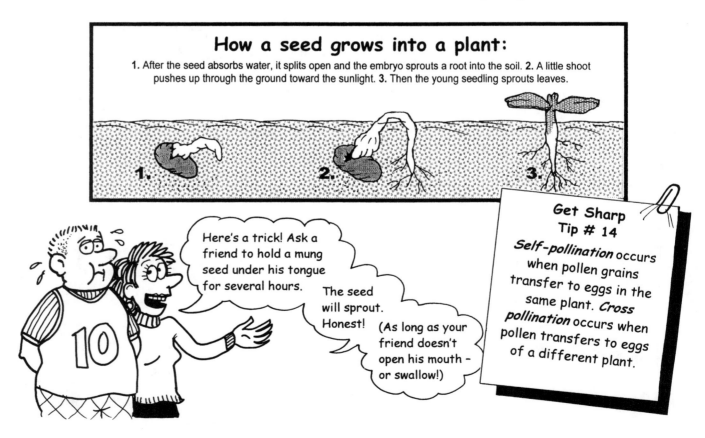

How a seed grows into a plant:

1. After the seed absorbs water, it splits open and the embryo sprouts a root into the soil. **2.** A little shoot pushes up through the ground toward the sunlight. **3.** Then the young seedling sprouts leaves.

Here's a trick! Ask a friend to hold a mung seed under his tongue for several hours.

The seed will sprout. Honest!

(As long as your friend doesn't open his mouth – or swallow!)

Get Sharp Tip # 14

Self-pollination occurs when pollen grains transfer to eggs in the same plant. *Cross pollination* occurs when pollen transfers to eggs of a different plant.

Reproduction in Flowering Plants

A flower is the reproductive part of an angiosperm. The male part of the flower is the *stamen*. It has an *anther* where pollen is formed, held up by a stalk called a *filament*. The female part is the *pistil*. It has an *ovary* where eggs are produced inside ovules, a *style* (stalk), and a *stigma,* which is the sticky top of the style.

Pollination occurs when pollen grains from the anther stick on the stigma. A pollen grain sends a tube down the style into an ovule within the ovary. Then a sperm, or male sex cell, travels down the pollen tube and joins with an egg. The result is a fertilized egg, or zygote, which grows into an embryo. The ovule around the zygote develops into a seed. The ovary around the seeds grows into a fruit. This offers protection to the seeds.

When the fruit dries up and withers, the seeds fall out onto the ground and grow into new plants.

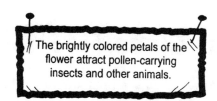

The brightly colored petals of the flower attract pollen-carrying insects and other animals.

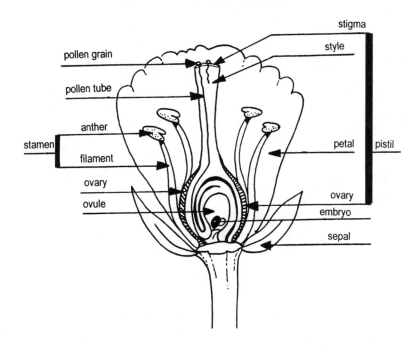

Reproduction in Conifers

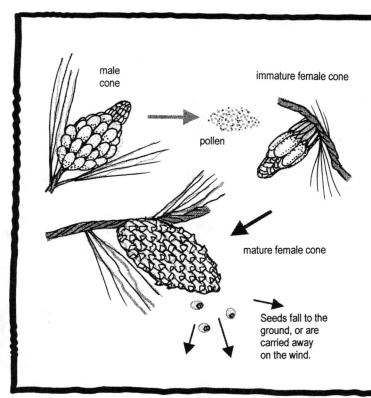

Seeds fall to the ground, or are carried away on the wind.

In a conifer, male cones produce large amounts of pollen. The pollen is released by the cones and carried by wind to female cones. Pollen grains fall on the female cone.

Each grain sends a tube down into an ovule within the female cone. Sperm then travels down the tube and fertilizes an egg in the ovule. A zygote, or fertilized egg, is formed.

The zygote grows into an embryo and the ovule develops into a seed. Later, the seeds fall from the female cones. Some of the seeds germinate and grow into new plants.

Animal Classification

There are more than a million different animal species, classified into more than 20 phyla. Animals within a phylum share common characteristics. Here are some of the largest phyla.

PORIFERA

This phylum contains animals known as sponges. Each organism is a thick sack of cells that form pores, chambers, or canals. They live in water and attach themselves to one place where they stay for life.

COELENTERATA

Coelenterates have a central cavity with a mouth. They live in water. Most have tentacles. Some stay attached to one place. *Examples: sea anemones, jellyfish, and coral.*

MOLLUSCA

Mollusks have soft bodies with hard coverings. They live on land or in water. Many of them have a thick, muscular foot for movement. Some live attached to surfaces such as rocks. *Examples: octopuses, snails, slugs, squids, and clams.*

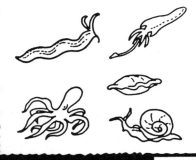

There are more than 15,000 species of nemotodes in the world.

Can you imagine that many worms?

PLAYTHELMINTHES

The common name is flatworms. These are flat-bodied worms that live as parasites or move freely in the water.

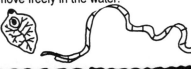

NEMOTODA

The common name is roundworms. These are round-bodied worms that live as parasites or move freely in water or on land.

ANNELIDA

The common name is segmented worms. These worms have bodies divided into segments with bristles. They live on land or in water. They are not parasitic.

ECHINODERMATA

Echinoderms have radial symmetry and a tough outer covering with spines. They all live in salt water and have a water vascular system for movement. They have tube feet which attach to objects and help them move. *Examples: starfish, sea urchins, sea cucumbers, and sea lilies.*

Should I put sea cucumber in my salad?

ARTHROPODA

Arthropods are animals with jointed limbs, and most have a hard, plated body covering. This is the largest animal phylum. These are the major classes in the phylum.

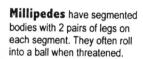

Insects have 3 body segments, 1 pair of antennae, and 3 pairs of legs. Many insects have wings. Flies, grasshoppers, beetles, bees, and butterflies are insects.

Arachnids, like spiders, scorpions, ticks, and mites, have an exoskeleton, segmented bodies, 2 body regions, 4 pairs of jointed legs, and no antennae. They have 2 pairs of jointed structures near the mouth used to hold and chew food.

Millipedes have segmented bodies with 2 pairs of legs on each segment. They often roll into a ball when threatened.

Centipedes have segmented bodies and 2 antennae on the head. There is a pair of legs on each body segment. The mouth has parts that bite and chew.

Crustaceans have segmented bodies with two regions. The head and thorax are joined; the abdomen is separate. They have claw-like legs at the end of their bodies, 2 pair of antennae, and sharp jaws for biting and chewing. Lobsters, shrimp, crayfish, and crabs are crustaceans.

CHORDATA

Chordates (or vertebrates) have internal skeletons made of bones or cartilage. They also have a central nervous system and specialized body systems for digestion, circulation, and respiration.

Amphibians are cold-blooded animals. Their skin is moist with no scales. They breathe air and live on land or water. They have 3-chambered hearts.

Fish are cold-blooded animals that live in water and breathe with gills. They have 2-chambered hearts, scaly coverings, and skeletons of bone or cartilage.

Reptiles are cold-blooded animals that breathe air and live mostly on land. Their bodies are scale-covered. They have 3-chambered hearts.

Mammals are warm-blooded animals that produce milk to feed their offspring. They are covered with hair or fur, have glands that produce sweat, and have 4-chambered hearts.

Birds are warm-blooded animals with wings and feather coverings. They have hollow, lightweight bones and hearts with 4 chambers.

ASK Dr. Phylla

Are spiders fast?

A spider can scurry along at over 1 mile per hour. That's faster than a turtle can move!

Which animal is the most numerous?

Insects are the most numerous animals on Earth. About 90% of all animals are insects.

Do centipedes have 100 legs?

Centi means 100. But not all centipedes have 100 legs.

Is a seahorse a swimming horse?

A seahorse is not really a horse. It's a fish!

Is a whale a fish, too?

The largest animal in the world, the blue whale, is a mammal. A blue whale can weigh up to 170 tons.

Get Sharp: Animal Classification

Animal Reproduction

In order to survive, a species must produce offspring to replace the members of the species that die off. Reproduction is the process that solves this problem. All animals reproduce. Here are some of the different methods for reproduction.

Asexual Reproduction

In *asexual reproduction*, there is only one parent and no combination of material from different sex cells.

Budding – Some animals reproduce by growing a new organism from a ***bud*** on the parent. A bulge of extra tissue grows on the parent animal. When the tissue is fully developed into an identical copy of the parent, the tissue breaks off, forming a separate offspring. Budding is common in some simple invertebrates such as sponges, hydras, and sea anemones.

BUDDING

Another me! Hi, Bud!

Hi, you.

Fragmentation – Sometimes an animal simply divides into two or more pieces. Each piece grows the missing parts and becomes a whole offspring. Fragmentation is common among many flatworms.

Parthenogenesis – Some insects, such as stick insects, produce offspring that grow from eggs which have not been fertilized.

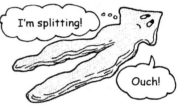

I'm splitting!

Ouch!

FRAGMENTATION

Papa!

Call me Daddy Dearest.

Seahorse fathers carry fertilized eggs in a pouch on their bodies until they are ready to hatch into tiny seahorses.

Sexual Reproduction

In *sexual reproduction*, material from two different sex cells combine to form a new individual. The joining of a ***sperm*** (male cell) with an ***egg*** (female cell) is called ***fertilization***. This process produces a ***zygote*** (fertilized egg) which then develops into a new individual.

External fertilization takes place outside the animal's body. Often a female releases or deposits eggs, then the male releases sperm which find their way to fertilize the eggs. Most fish, amphibians, and mollusks use this method of fertilization.

Internal fertilization takes place inside the female animal's body. The sperm is deposited in the female's body during mating. In some animals, such as reptiles, birds, and many arthropods, the eggs are deposited and hatched outside the animal's body. In most mammals and some reptiles, the eggs develop inside the female's body. The new animal is born after it is fully developed.

Metamorphosis

There are many species of animals that reproduce sexually but produce offspring that do not have all the same structures as the adult animal. These animals go through a series of changes or stages before taking the shape of the complete adult.

Complete Metamorphosis

In *complete metamorphosis*, the newborn offspring has little resemblance to the adult animal. Butterflies, bees, ants, flies, and moths all undergo complete metamorphosis.

BUTTERFLY	FROG

The butterfly begins with a fertilized egg, or **zygote**. The zygote hatches into a **larva.** (A butterfly larva

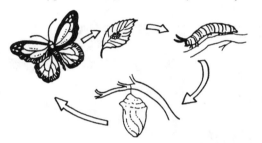

is called a **caterpillar**.) The larva eats and grows, shedding its skin several times. When its growth is finished, it forms a hard case called a **pupa.** (A butterfly pupa is called a **chrysalis**.) The butterfly develops inside the pupa, and hatches from it.

The frog's eggs develop into **tadpoles**. A tadpole breathes with gills and eats plants found in the pond.

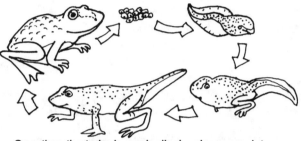

Over time the tadpole gradually develops complete gills and lungs, and grows legs. As an adult, the frog is able to breath on land, while the tadpole can only survive in the water. The tail is absorbed into the body as the little frog grows.

Incomplete Metamorphosis

Incomplete metamorphosis involves three stages of development. The newborn offspring looks like a small adult but does not have all the structures of an adult.

GRASSHOPPER

The tiny grasshopper that hatches from a fertilized egg is called a **nymph**. It looks like a miniature adult grasshopper,

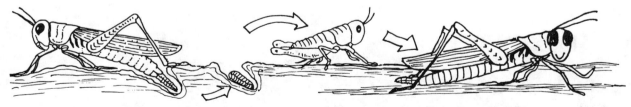

except that it has no wings or reproductive organs. As it grows, it sheds its **exoskeleton** (its hard, outer shell) several times and gradually develops wings and reproductive organs.

155

Animal Behavior

Now and then, a darkling beetle will stand on his head. Others beetles roll up into balls. Fiddler crabs change color twice a day. Hognose snakes play dead. Flamingos stand on one leg. African grasshoppers blow bubbles. A daddy frog holds baby tadpoles inside his mouth for weeks. Monarch butterflies fly 4,000 miles to Mexico for a winter vacation. These are just some of the fascinating activities that are part of animal behavior.

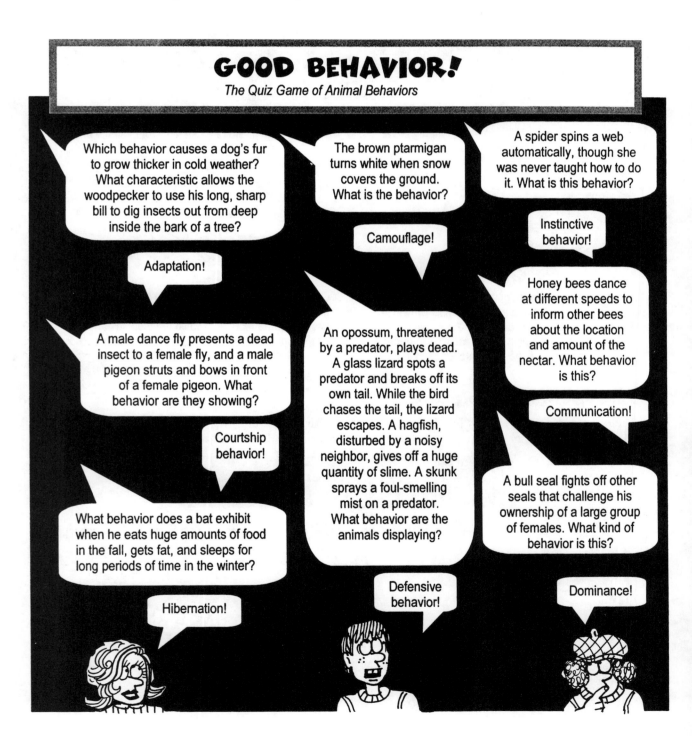

GOOD BEHAVIOR!
The Quiz Game of Animal Behaviors

Which behavior causes a dog's fur to grow thicker in cold weather? What characteristic allows the woodpecker to use his long, sharp bill to dig insects out from deep inside the bark of a tree?

Adaptation!

The brown ptarmigan turns white when snow covers the ground. What is the behavior?

Camouflage!

A spider spins a web automatically, though she was never taught how to do it. What is this behavior?

Instinctive behavior!

A male dance fly presents a dead insect to a female fly, and a male pigeon struts and bows in front of a female pigeon. What behavior are they showing?

Courtship behavior!

An opossum, threatened by a predator, plays dead. A glass lizard spots a predator and breaks off its own tail. While the bird chases the tail, the lizard escapes. A hagfish, disturbed by a noisy neighbor, gives off a huge quantity of slime. A skunk sprays a foul-smelling mist on a predator. What behavior are the animals displaying?

Honey bees dance at different speeds to inform other bees about the location and amount of the nectar. What behavior is this?

Communication!

What behavior does a bat exhibit when he eats huge amounts of food in the fall, gets fat, and sleeps for long periods of time in the winter?

Hibernation!

Defensive behavior!

A bull seal fights off other seals that challenge his ownership of a large group of females. What kind of behavior is this?

Dominance!

ANIMAL BEHAVIOR CATEGORIES

ADAPTATION characteristics or changes that enable an animal to survive in its environment	CAMOUFLAGE changing body color or appearance for protection	COMMUNICATION giving messages, warnings, or other information to each other	COURTSHIP BEHAVIOR actions surrounding the mating process
DEFENSIVE BEHAVIOR actions that help the animal protect itself from danger	DOMINANCE behavior that shows or keeps an animal's status or power over other animals	REGENERATION growing a new body part after one is broken off	INSTINCTIVE BEHAVIOR inborn responses to stimuli; reactions that were not taught
MIGRATION moving long distances to reproduce, mate, raise young, or find food	MIMICRY copying the appearance or behavior of something else for protection	HIBERNATION a long period of rest or inactivity, usually in the winter	NURTURING BEHAVIOR actions of taking care of offspring
REFLEX BEHAVIOR automatic response to stimulus	SOCIAL BEHAVIOR animals living in groups for survival or company	TERRITORIALITY actions designed to protect or defend a certain geographical area	MOVEMENT getting from place to place

Blue whales spend the summer in the polar ocean where there is plenty of food, and in the fall, they swim south to warm waters near the equator. The arctic tern travels 22,000 miles round trip to its summer home. Can you name this behavior?

Migration!

A male King Penguin keeps a new-born chick safe and warm under a flap of skin on his feet. Name this behavior.

Nurturing behavior!

A deer, munching leaves in the forest, jerks her head up when a twig snaps nearby. Why?

Reflex behavior!

The viceroy butterfly fools predators by looking just like the foul-tasting monarch. An anglerfish looks just like the rocks on the ocean floor. Why?

Mimicry!

What behavior do wolves exhibit when they live and hunt in packs for success?

A goose hisses at any animal that comes into his home area. A wolf urinates around the edge of area where he lives. What is the behavior?

Territoriality, of course!

A crab loses a claw in a fight. In a few days he grows a new one. How?

Social behavior!

Regeneration!

A snail moves in an unusual way. Rhythmic waves flow through the muscular foot, pushing it slowly forward or drawing it back. The snail produces a slippery slime, which helps the foot move easily. Name this behavior.

Movement!

Get Sharp: Animal Behavior

Biomes

Every plant or animal lives in a biome. What is a **biome**?
It's a region with a distinct climate, a dominant type of
plant, and specific organisms that are characteristic
of the region. There are several different biomes.
Each one is home to many organisms.

Fresh Water Biome

Characteristics:
- found in rivers, streams, lakes, ponds, swamps, marshes
- rich in plant & animal life living in and near the water

Plant & animal life:
green algae, pond weed, flowers, cattails, fish, crayfish, snakes, turtles, birds, alligators, frogs, insects

Salt Water Biome

Characteristics:
- found in Earth's oceans and seas
- organisms are adapted to salt water

Plant & animal life:
algae, phytoplankton, fish, seals, whales, sponges, mollusks, coelenterates, and echinoderms

Tropical Rain Forest Biome

Characteristics:
- found near equator precipitation—over 100 in. annually
- consistently hot climate — average 80° F temperatures

Plant & animal life:
tall evergreen trees, fruit trees, vines, thick leafy plants, birds, monkeys, leopards, amphibians, snakes, insects, jaguars

Grassland Biome

Characteristics:
- rich soil
- 10-40 in. of rain annually
- hot, dry climate
- herds of grazing animals

Plant & animal life:
thick grasses (some very tall), gazelles, zebras, wildebeests, bison, lions, hyenas, vultures, small burrowing animals

Desert Biome

Characteristics:
- hot temperatures, cool nights
- less than 10 in. rain annually
- organisms adapted to limited water
- little plant life

Plant & animal life:
succulent plants that store water, lizards, snakes, rabbits, mice, insects, birds, camels

Temperate Deciduous Forest Biome

Characteristics:
- found in temperate climates
- 40 in. of precipitation a year
- extensive deciduous forests
- many forests have been cleared for farmland or homes

Plant & animal life:
mixed deciduous trees, small mammals, deer, birds, rodents, foxes, insects, wildflowers

Taiga Biome

Characteristics:
- found in cold climates in Northern coniferous forests
- long, harsh winters
- damp ground, fog

Plant & animal life:
coniferous trees such as spruce, pine, and fir, bears, wolves, beavers, small mammals, insects

Tundra Biome

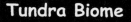

Characteristics:
- found near polar ice caps
- constant low temperatures
- permafrost
- land is wet and swampy in summer, frozen in winter

Plant & animal life:
mosses, grasses, lichen, birds, polar bears, wolves, caribou, walruses, mosquitoes, penguins

Get Smart Tip # 16
A biome does not have clear boundaries. Biomes overlap and characteristics change gradually between biomes.

159

Relationships in the Environment

Ecology is the study of the relationships between living things and their surroundings. The area that surrounds an organism, and all the other organisms within that area is the organism's *environment*. The particular place within the environment where an organism lives is called its *habitat*. All the plants and animals that live together in a habitat are a *community*. The word *ecosystem* is used to describe a community, its habitat, and all of the relationships within that habitat. Ecosystems can be as small as a puddle or a rotten tree stump, or as large as an ocean or a major desert. Every habitat has a limited amount of resources (space, sunlight, nutrients, soil, food, and water). Within an ecosystem, different species make use of different resources, so a delicate balance is formed that allows all the organisms to survive. When any one factor in the ecosystem changes or disappears, the balance may be disturbed.

The number of individuals of one species in a community is a *population*. In order to describe a community, it is helpful to understand the number and sizes of the different populations. Within an ecosystem, each species has a particular role, called a *niche*. The niche of a species includes the way that organism interacts with all the other organisms in the ecosystem.

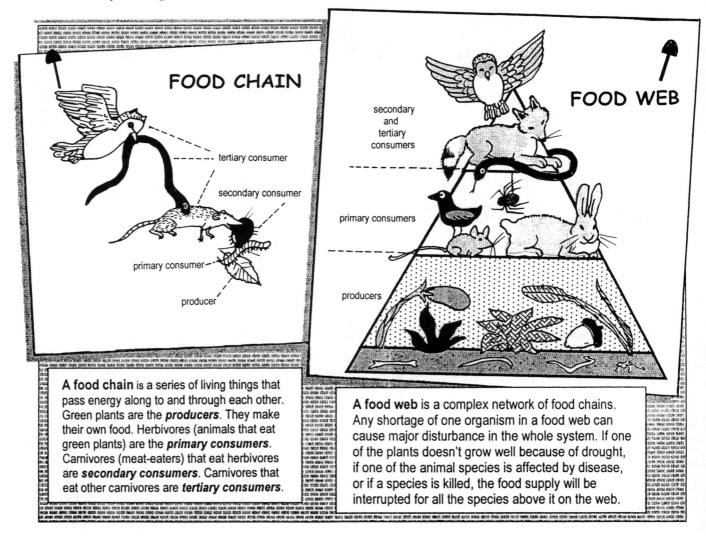

FOOD CHAIN

tertiary consumer

secondary consumer

primary consumer

producer

FOOD WEB

secondary
and
tertiary
consumers

primary consumers

producers

A food chain is a series of living things that pass energy along to and through each other. Green plants are the *producers*. They make their own food. Herbivores (animals that eat green plants) are the *primary consumers*. Carnivores (meat-eaters) that eat herbivores are *secondary consumers*. Carnivores that eat other carnivores are *tertiary consumers*.

A food web is a complex network of food chains. Any shortage of one organism in a food web can cause major disturbance in the whole system. If one of the plants doesn't grow well because of drought, if one of the animal species is affected by disease, or if a species is killed, the food supply will be interrupted for all the species above it on the web.

The organisms within an ecosystem interact with each other in a complicated series of relationships. These are some of the major relationships going on in an ecosystem.

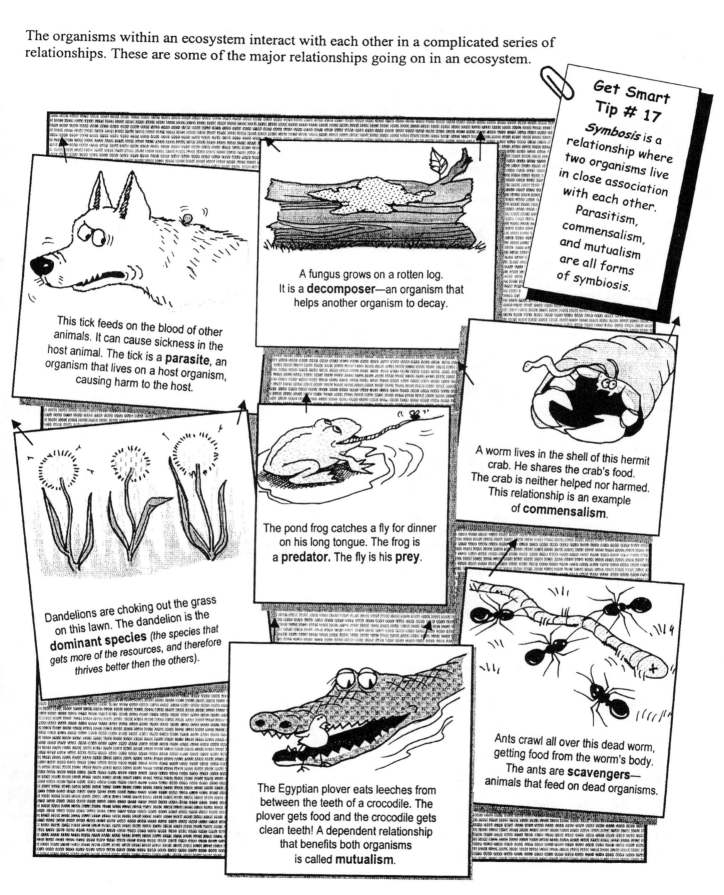

This tick feeds on the blood of other animals. It can cause sickness in the host animal. The tick is a **parasite**, an organism that lives on a host organism, causing harm to the host.

A fungus grows on a rotten log. It is a **decomposer**—an organism that helps another organism to decay.

A worm lives in the shell of this hermit crab. He shares the crab's food. The crab is neither helped nor harmed. This relationship is an example of **commensalism**.

Dandelions are choking out the grass on this lawn. The dandelion is the **dominant species** (the species that gets more of the resources, and therefore thrives better then the others).

The pond frog catches a fly for dinner on his long tongue. The frog is a **predator.** The fly is his **prey**.

Ants crawl all over this dead worm, getting food from the worm's body. The ants are **scavengers**— animals that feed on dead organisms.

The Egyptian plover eats leeches from between the teeth of a crocodile. The plover gets food and the crocodile gets clean teeth! A dependent relationship that benefits both organisms is called **mutualism**.

Get Sharp: Relationships in the Environment

Problems in the Environment

The Earth has limited resources, such as water, air, and fuels. The living things on Earth make great demand on those resources. Some living things (such as people) make use the environment's resources more heavily than others. Some of the resources are running out. Others are becoming damaged or polluted. *Pollution* is the release of harmful substances into the environment. Every day tons of poisonous gases, chemicals, sewage, and garbage are poured into the water, soil, and air. These substances harm or kill organisms, and destroy the delicate balance in ecosystems. In addition to substances that pollute, the environment is polluted with noise, unattractive sights, and excess heat.

Some Terms & Concepts to Know

Acid rain (or snow) is created when sulfur dioxide, released into the air by industries, combines with water vapor in the air. It is harmful to plants, animals, metals, and stone.

Ozone layer is a thick layer about 15 miles up in the atmosphere. Ozone is a form of oxygen that protects the Earth from the Sun's harmful ultraviolet rays. Polluting chemicals called *chlorofluorocarbons* are destroying or "eating a hole" in this protective layer.

Greenhouse gases such as carbon dioxide are trapped close to the Earth by the atmosphere. As levels of these gases increase, the Earth gets warmer. Greenhouse gases result from the burning of fossil fuels.

Global warming is the term used to describe the gradual increase in temperatures around the world. This may cause excess melting of the polar caps, which could result in serious flooding.

Deforestation is the removal of forests, usually to convert land to farms or use them for development. Deforestation can lead to erosion and loss of plant and animal species. As trees are cut, there is less plant life to use carbon dioxide in photosynthesis. This adds to the *greenhouse effect*.

Smog is polluted fog. Smog is particularly harmful to the respiratory system. It also blocks sunlight.

Fossil fuels are fuels, such as coal, petroleum, and natural gas formed from layers of organisms that have decayed beneath Earth's surface. These are nonrenewable resources that take millions of years to form.

Renewable resources are resources that can be replaced by nature over a period of time (crops, land, trees, plants, animals).

Nonrenewable resources are resources that take hundreds to millions of years to form (coal, petroleum, natural gas). The supply is limited, but with careful use the supply can last longer.

Sewage is human waste material. Untreated sewage is often dumped into oceans, lakes, or rivers. As it decays, it can be fatal to living things.

Thermal pollution is the raising of temperature in a river or other body of water when hot water is dumped by industries. The higher temperatures cause growth of organisms that harm fish.

Erosion is the removal or movement of Earth materials (such as soil).

SOURCES OF HARM
TO THE ENVIRONMENT

FACTORY FUMES
SEWAGE AND GARBAGE
PERSONAL SEWAGE
AND GARBAGE
SPRAYED SUBSTANCES
VEHICLE EXHAUST
BURNING FUELS
PESTICIDES AND FERTILIZERS
OIL SPILLS
SHIPS DUMPING WASTES
MINING
DEFORESTATION
ACID RAIN
NATURAL DISASTERS SUCH AS
DROUGHTS,
FLOODS, AND STORMS

Help for the Troubled Earth

Many methods, programs, changes, and practices are being tried to reduce harm and repair damage to the environment. These are a few of them.

Conservation is a term that applies to any efforts made to protect, replace, or make careful use of resources.

Soil conservation includes a variety of practices used to reduce erosion and improve fertility of soil. Some of these methods are uses of mulch, planting cover crops to hold soil, planting trees as windbreaks, contour planting and plowing, and crop rotation.

Reforestation is the practice of planting seeds or small trees to replace forests.

Forest management involves reducing waste in cutting and using timber, and managing forests to prevent fires, control diseases, reduce erosion, and keep healthy forests.

Use of renewable resources such as wind power and solar power slows the use of nonrenewable resources.

Wildlife preservation is the practice of maintaining any living species to protect them from extinction. This usually involves restoring and protecting wildlife habitats.

Reserves are areas of land that are protected for preserving the habitats of plants and animals.

Landscape preservation is the reclaiming of land that has been damaged by mining, industry, or dumping.

Natural pest control is the practice of using harmless and natural substances instead of chemical pesticides to control pest damage to crops.

Recycling is the practice of using things over and over again instead of throwing them away. Paper, aluminum, and glass are easily recycled.

Emissions Control is the practice of trying to reduce the amount of harmful emissions released into the environment by automobiles.

National and international agreements and organizations, such as the UN Earth Summits, Environmental Protection Agency, World Wildlife Fund, the Sierra Club, and National Resources Conservation Services, have different goals and responsibilities related to preserving Earth's resources.

Recycling is a practice that anyone can do to help the environment. Make it a habit!

Some Websites to Explore
Environmental Protection Agency: www.epa.gov
World Wildlife Fund: www.worldwildlife.org
Conservation Kids: www.kidsgowild.com
Conservation International: www.conservation.org
National Resources Conservation Service: www.nres.usda.gov

163

How the Body Works

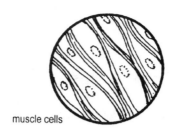
muscle cells

The body is a complex machine of different parts. Each part has an important job to do, and the parts work together as systems to move, sense the world, communicate, breathe, reproduce, get and use energy. If all the body parts are working properly, they function together to keep a healthy, working body.

Cells

The cell is the body's basic unit of life. There are different kinds and shapes of cells in the body, but most body cells are similar in structure. (See the animal cell on page 139.) There are more than 50 billion cells in the human body.

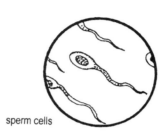
sperm cells

Tissues

A *tissue* is a group of cells that have a similar shape and function. There are four different kinds of body tissues. Many organs are composed of more than one type of tissue.

Epithelial tissue (skin tissue) forms the outer layer of skin and lines the inner surfaces of the blood vessels and digestive tract.

Connective tissue surrounds organs and connects tissues together. Blood, bone, and cartilage are connective tissues.

Nerve tissue is made up of nerve cells that are able to send signals around the body.

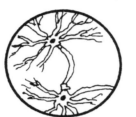

Muscle tissue is made of cells that can contract. Muscle tissue is found in the muscles that attach to the skeletal structures, in the heart, and in other places such as the intestines and stomach.

nerve cells

Organs

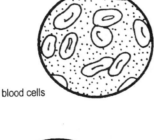
blood cells

An *organ* is a group of different tissues that work together to perform a certain function in the body. The heart, for instance, is a group of tissues such as muscle and nerve tissue. These work together to keep the heart doing its job of pumping blood around the body.

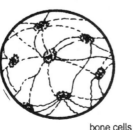
bone cells

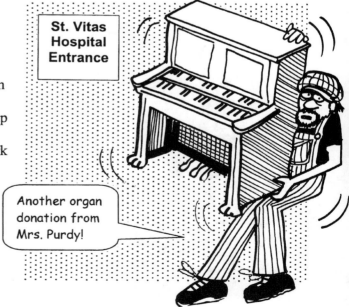

St. Vitas Hospital Entrance

Another organ donation from Mrs. Purdy!

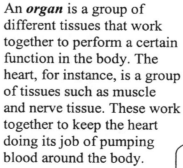

164

Respiratory System

Cells in the body must have oxygen in order to work. Oxygen is needed for cells to release energy from food. The *respiratory system* is a group of organs that takes in oxygen from air, transfers the oxygen to the blood, removes wastes (carbon dioxide) from the blood, and expels the wastes from the body.

The larynx is the voice box, which contains your vocal cords.

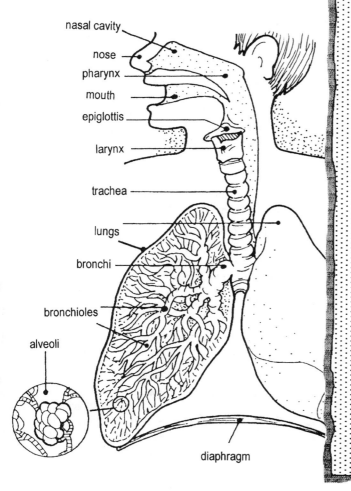

nasal cavity
nose
pharynx
mouth
epiglottis
larynx
trachea
lungs
bronchi
bronchioles
alveoli
diaphragm

Breathing

1. As you breathe in, dust and germs are caught by hairs in the nose and mucous in the **nose and throat**. The air travels through the mouth into the **pharynx** (throat), down the **trachea** into the left and right **bronchi** (structures within the lungs). It continues into the smaller **bronchioles** and the tiny **alveoli** (air sacs) located at the ends of the bronchioles.

2. The alveoli are surrounded by tiny blood vessels called **capillaries**. Oxygen passes through the walls of the alveoli into the capillary cells. The capillaries and other blood vessels carry **oxygen-rich blood** throughout the body.

3. The blood picks up carbon dioxide wastes from all the body's cells. This is brought back to the capillaries, and transferred through the walls of the alveoli into the lungs. The **carbon-dioxide laden air** is breathed out, traveling up from the lungs through the bronchioles, bronchi, trachea, **pharynx** (throat), and out the mouth or nose.

The **diaphragm** is a muscle that helps with breathing. When you inhale, it expands downward, making the chest cavity bigger and allowing the lungs to fill with air. When you exhale, the diaphragm contracts and pushes up, making less room for air and helping to force air out of the body.

Get Sharp: Body Systems

Digestive System

Before the food you eat can provide energy for the body's cells, it needs to be changed and processed into a usable form. This process is called *digestion*. The digestive tract is a long tube that twists and turns its way from the mouth to the anus.

The taco that I ate last night is on a 24-hour journey through a 30-foot long tube. On its way, most of it will be broken down, mashed, and dissolved and changed into a substance that can be absorbed by my cells.

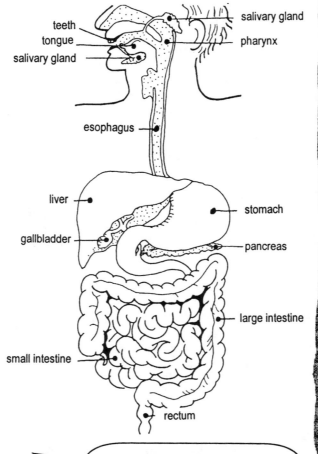

teeth
tongue
salivary gland
salivary gland
pharynx
esophagus
liver
gallbladder
stomach
pancreas
large intestine
small intestine
rectum

In 1927, doctors examined a woman complaining of intestinal discomfort. It turns out that she had swallowed 2533 objects, all lodged in her stomach. This included 947 bent pins.

Food's Journey

1. ***Teeth*** bite the food and chew it into tiny pieces. ***Saliva*** made by the ***salivary glands*** helps to dissolve the food with an ***enzyme*** (chemical that changes food).

2. Muscles in the ***tongue*** push the food into the ***pharynx*** (throat). Throat muscles help you swallow the food. As you swallow, the ***epiglottis*** closes over the trachea, so food won't go towards the lungs.

3. The food travels into a long tube called the ***esophagus***, which has muscles that squeeze the food along to the stomach.

4. The ***stomach*** is a powerful muscle. It squeezes and churns food for about 4 hours, mixing in enzymes which begin digesting proteins. ***Hydrochloric acid*** in the stomach kills any bacteria in the food.

5. The ***liver*** makes a green liquid called ***bile*** whose job it is to break up fats. Bile is stored in your ***gallbladder***. The ***pancreas*** is a gland that makes digestive enzymes also.

6. When food leaves the stomach, it goes into the first part of the ***small intestine*** (the ***duodenum***) where bile and pancreatic juices mix with the food and digest it. The second half of the small intestine (the ***ilium***), is lined with thousands of tiny finger-like structures called ***villi***. The walls of the villi are one cell. Digested food passes through the walls of the villi into the tiny capillaries within the villi.

7. Undigested food and water move along into the ***large intestine***. The water passes though the walls of the first part of the large intestine (the ***colon***). Solid waste matter is stored in the last part of the large intestine (the ***rectum***) until the muscles push it out though the ***anus***.

Better Grades & Higher Test Scores / SCIENCE
Copyright ©2003 by Incentive Publications, Inc., Nashville, TN.

Waste Removal Systems

Waste products are created as a part of many body processes. These can be poisonous to the cells of the body and cause serious damage or death to body tissues. *Excretion* is the process of getting rid of waste products. The body has several ways to get rid of them.

That double chocolate caramel sundae went right through me!

What a waste!

The digestive system removes undigested solid wastes and water in the large intestine.

The lungs get rid of carbon dioxide and water vapor.

The skin is an organ that removes excess water from the body (as *sweat*) to help regulate body temperature. The sweat also contains small amounts of waste mineral material.

Kidneys filter the blood, removing water, mineral, and protein wastes. A liquid called *urine* is formed from the filtered wastes. Urine is removed from the kidneys by two long tubes called *ureters,* which carry the urine to the *bladder,* a muscular storage bag. Contractions of the bladder expel urine out of the body through a tube called the *urethra*.

The liver filters harmful chemical substances from the blood.

Endocrine System

Chemical substances called *hormones* control many processes in the body. Special tissues called *glands* produce the hormones. The hormones are carried throughout the body by the blood. Different glands produce different hormones, each of which has a specific function.

My mom says that teenagers have raging hormones.

I wonder what they're mad about?

Gland	Gland Location	Hormone Produced	Function of the Gland and Hormone
pituitary	near the brain	11 different hormones	the "master gland'; controls growth; controls many other glands
adrenals	near kidneys	adrenalin	regulates metabolism and body activity in emergencies
pancreas	below & behind stomach	insulin	controls amount of sugar in the blood and storage of sugar in the liver
parathyroid	behind thyroid	parathormone	regulates calcium and phosphorus in the blood and tissues
thyroid	neck	thyroxine	controls metabolism in the body (the rate at which the body uses its food)
ovaries	female pelvic area	estrogen	controls the production of eggs; controls female characteristics
testes	male pelvic area	testosterone	controls the production of sperm cells; controls male characteristics

---

167

Skeletal-Muscular System

The human skeleton has more than 200 bones. Along with the muscles, these give strength and shape to the body. They also provide a protection for the fragile organs inside the bones, and they are a great place for muscles to attach so the body can move.

Joints occur at places where bones meet. They allow the body to bend and twist. Most joints are padded with a rubbery tissue called **cartilage**. Bones are held together at joints by strong, flexible bands called **ligaments**. There are several kinds of joints. A **ball and socket joint** allows the shoulder and hip to swivel. A **hinge joint** allows the elbow and knee to bend. A **sliding joint** allows the spine to twist and bend. A **pivot joint** allows the wrist to twist.

Major Bones

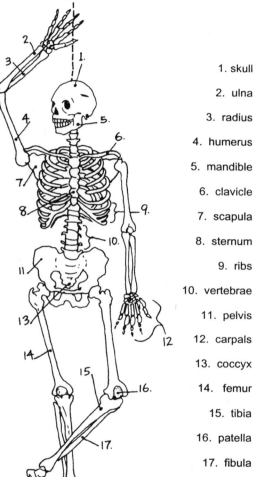

1. skull
2. ulna
3. radius
4. humerus
5. mandible
6. clavicle
7. scapula
8. sternum
9. ribs
10. vertebrae
11. pelvis
12. carpals
13. coccyx
14. femur
15. tibia
16. patella
17. fibula

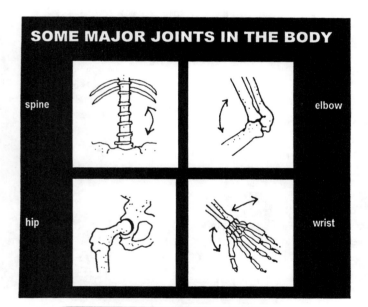

SOME MAJOR JOINTS IN THE BODY

spine

elbow

hip

wrist

According to the *Guinness Book of Records*, the biggest feet of any living person belonged to Matthew McRory of Pennsylvania. He wore size $28\frac{1}{2}$ shoes! It's no surprise that his mother had to knit his socks!

Wow! Now there's some long bones!

Muscle Talk

Skeletal muscles generally work in pairs. They pull or relax. They never push. Look! When at what happens when my biceps muscle contracts to pull my arm up! It bulges!

When the biceps contracts, the triceps muscle is relaxed.

When I straighten my arm, the biceps relaxes and the triceps contracts to pull the arm down.

Muscle tissue is used to move all parts of your body. It even helps the beating of your heart and the movement of food through the digestive system. ***Voluntary muscles*** move when you choose for them to move. ***Involuntary muscles***, like the heart muscle, muscles in the intestines, or muscles that blink your eyes, move automatically.

Striated muscles are voluntary muscles. These are also known as ***skeletal muscles*** because they move the bones of the skeletal system. They're called striated because they look striped. These muscles are attached to the bones with tendons.

Smooth muscles line the insides of many body organs. Smooth muscles are involuntary.

Cardiac muscle has only one location—that's in your heart. It is an automatic, or involuntary, muscle.

The biggest biceps in the world belongs to a man who started building his muscles wrestling pigs. Denis Sester's biceps measures over 30 inches.

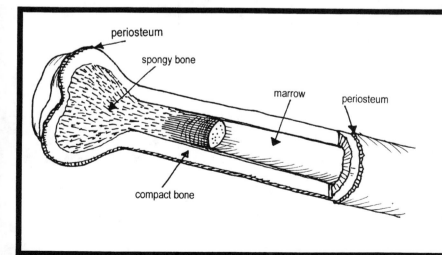

periosteum
spongy bone
marrow
periosteum
compact bone

A look inside a bone

The outer layer of the bone is a very thin, tough layer, called the ***periosteum***. When a bone breaks, the periosteum cells multiply and grow over the break. Hard ***compact bone*** lies beneath the periosteum. The inner bone is ***spongy bone***. It is light, with lots of holes, but it is very strong. Bone gets its strength and hardness from phosphorus, calcium, and a protein material called ***collagen***. Many bones have a soft, inner tissue called ***marrow.*** Red blood cells are made in the ***red marrow***.

169

Nervous System

The **nervous system** is a busy and complex network of nerves. Each nerve is a bundle of nerve fibers that extend from nerve cells or **neurons**. Most of the bodies of the nerve cells are located in the brain or spinal cord. The **brain** and the **spinal cord** are the **central nervous system**. All of the rest of the nerves make up the **peripheral nervous system** and carry messages between the central nervous system and the rest of the body. The nerves send signals to and from the brain, and control **autonomic** (automatic) processes such as digestion, circulation, and breathing.

Neurons

Neurons (D) are cells that carry electric signals (**impulses**). A neuron has a cell body with a nucleus **(F)** and **dendrites** (long branches) **(C)**, and **axons** long fibers) **(B)**. Some have **receptors (A)**. Receptors in sensory neurons receive a stimulus and pass it along the axons to the dendrites of connector cells. Impulses travel across **synapses** (small spaces between cells) **(E)** until they reach motor neurons.

Sensory neurons carry impulses from receptors in sense organs to the central nervous system, **motor neurons** carry impulses away from the central nervous system to muscles and glands, and **interneurons** connect sensory neurons and motor neurons.

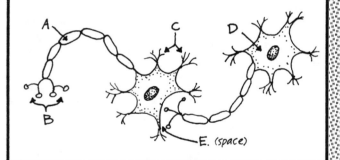

The Brain

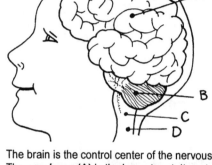

The brain has about 10 billion cells!

The brain is the control center of the nervous system. The **cerebrum (A)** is the largest part. It controls thinking, learning, memory, awareness, and some voluntary movements. It also controls all senses. At the back of the head lies the **cerebellum (B)**. It controls muscle activity and balance. The **medulla (C)** is the smallest part of the brain. It lies at the base of the skull. It controls automatic functions such as breathing, the heartbeat, gland secretions, reflexes, and digestive action. The **spinal cord (D)** is a thick cord of nerves that runs through the vertebrae. It carries messages between the brain and the rest of the body.

Touch

The skin is a sensory organ that has many nerve endings. It is sensitive to touch, heat, cold, and pain. The top layer, the **epidermis (A)**, keeps out germs. The second layer, the **dermis (B)**, is thicker. It contains the **nerve endings (C)**, **sweat glands (D)**, **blood vessels (E)**, and **hair roots (F)**. Sweat glands help to regulate the body's temperature by passing sweat out of the **pores (G)**. As the sweat evaporates, the body cools.

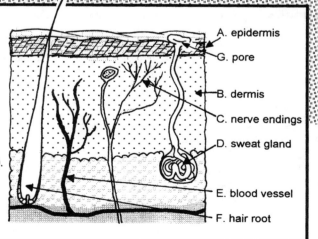

A. epidermis
G. pore
B. dermis
C. nerve endings
D. sweat gland
E. blood vessel
F. hair root

The skin on your fingers, lips, and the soles of your feet have more sensory neurons than anywhere else.

Sight

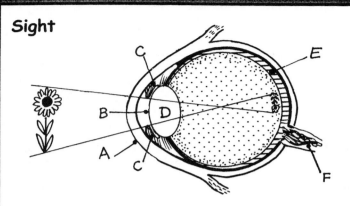

Every image constantly reflects rays of light. Carrying the reflected image, the rays pass through the **cornea (A)**, the eye's thin protective outer layer, and enter into the eye through the **pupil (B)**. The pupil is a small hole in the center of the eye, surrounded by the **iris (C)**, a muscle that controls the size of the hole. The light passes through a transparent disc called a **lens (D)**, which bends the light to focus it on the retina. The image is upside down when it reaches the retina. The **retina (E)** is the "screen," an area in the back of the eye that contains receptor cells. These cells send impulses on through the **optic nerve (F)** to the brain, which interprets the signals. The brain turns the image right side up.

Hearing

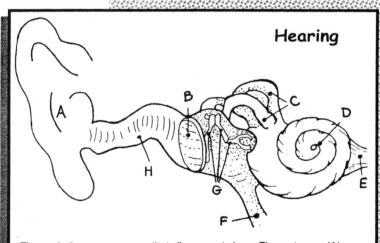

The ear is the sensory organ that allows you to hear. The **outer ear (A)** catches sounds and directs them into the ear, down a tube called the **auditory canal (H),** which is lined with hairs and produces wax. The **eardrum (B)**, a stretched membrane, separates the outer ear from the **middle ear**. The eardrum vibrates when sound hits it. Three tiny bones in the inner ear, the **hammer, anvil,** and **stirrup (G)**, pick up vibrations and pass them along to the **inner ear**. The **cochlea (D),** a coiled liquid-filled tube, picks up vibrations from the middle ear and passes them along to the **auditory nerve (E)**. This nerve sends sound signals to the brain.

The **semicircular canals (C)** in the inner ear help to maintain balance while the body is in motion. The **Eustachian tube (F)** leads from the back of the nose to the ear. This is the only way air can get in and out of the middle ear to keep pressure equal on both sides of the eardrum.

Smell

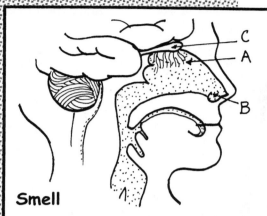

The nose is the organ responsible for the sense of smell. **Scent-sensitive cells (A)** in the back of the nose respond to scents dissolved in **mucous** in the **nose (B)**. Signals are carried from those cells by other nerve cells to the brain's **olfactory lobe (C)**. This is the area where smells are recognized.

Taste

Taste is closely related to smell. The sense of smell is actually much stronger than the sense of taste, so tasting relies partly on the sense of smell. **Taste buds** at the front, sides, and back of the tongue are groups of receptor cells that are sensitive to chemicals dissolved in the saliva. Buds in different areas of the tongue respond to different tastes.

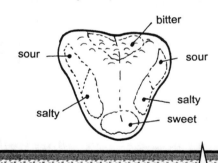

In 1978, a man ate 91 pickled onions in just over 1 minute.

Which of the 10,000 taste buds allow a person to taste a pickled onion?

Circulatory System

The *circulatory system* is a vast network of large and small tubes called *vessels* that carry blood around the body. The blood transports valuable supplies of oxygen and food to all the body's cells and carries wastes away from the cells.

Blood

Blood consists of a mixture of different kinds of blood cells floating in a yellow liquid. Blood cells are made inside the marrow of bones.

Red blood cells carry oxygen. As blood passes through the lungs, oxygen combines with a chemical compound called *hemoglobin* that is contained in the red blood cells.

White blood cells defend the body by producing antibodies to fight off disease and by engulfing harmful bacteria.

Platelets are fragments of blood cells that stop the bleeding when a vessel is damaged.

Plasma is the liquid that carries blood cells and platelets through the vessels.

Blood Vessels

Blood flows out from the heart through vessels that divide into smaller and smaller vessels down to microscopic capillaries surrounding body tissues and organs. The capillaries join together again to form larger and larger vessels carrying blood back to the heart.

Arteries are the vessels that carry blood away from the heart. They have thick walls because the heart pulses blood through them at a high pressure. The *aorta*, the main artery leaving the heart, is the body's largest artery. The *carotid artery* supplies blood to the brain.

Veins are the vessels that carry blood back to the heart. The walls of veins are thinner because the pressure is lower than in arteries. There are valves in veins to keep the blood from flowing backwards. The *superior vena cava,* coming into the heart, is the largest vein in the body.

Capillaries are the tiniest vessels, with walls one cell thick. Substances pass between the capillaries and the cells, delivering food and oxygen, and carrying away wastes.

St. Vitas Hospital Newspaper   January 12

Dr. Drake Q. Lah Joins Hospital Staff

The renowned researcher from Transylvania will be the new chief of hematology.

"I am very interested in all types of blood," said Dr. Lah. "Particularly when it's fresh."

Dr. Lah is looking for blood donors!

Blood Types

Everybody's blood falls into one of four groups: Type A, B, AB, or O. For a blood donation, the donor and patient must be matched carefully, because some blood contains antibodies against other blood types.

A can receive O, A
B can receive O, B
AB can receive all types
O can receive O

can donate to A, AB
can donate to B, AB
can donate to AB
can donate to all types

The Heart

The **heart** is the pump that makes the whole circulatory system work. It is a strong muscle that pumps blood through a network of almost 70,000 miles of blood vessels. The heart is divided into four **chambers**. Each side of the heart has an upper chamber (an **atrium**) and a lower chamber (a **ventricle**). As the blood flows through the heart, **valves** open and close between each atrium and ventricle, and between the ventricles and the arteries leaving the heart. The valves keep the blood from flowing backwards.

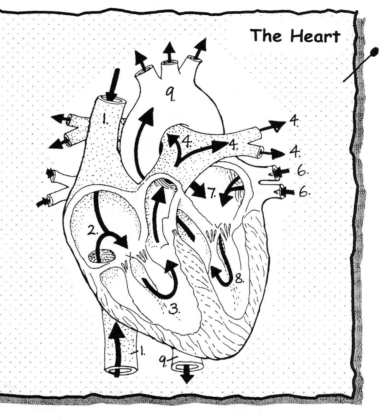

The Heart

1. The largest vein (the **superior vena cava**) brings blood that is low in oxygen and high in carbon dioxide into the heart.
2. The blood enters the **right atrium**.
3. The right atrium contracts and forces blood through a valve into the **right ventricle**.
4. The right ventricle contracts and forces blood through the **pulmonary arteries** toward the lungs.
5. In the lungs, the carbon dioxide is exchanged for oxygen.
6. This oxygen-rich blood flows back from the lungs through the **pulmonary veins**.
7. The oxygen-rich blood from the lungs enters the **left atrium**.
8. The left atrium contracts and forces blood through a valve into the **left ventricle**.
9. From the left ventricle, the heart pumps the blood out through the **aorta** into the body.

Ask Dr. A Orta

Dear Dr. A. Orta,
What causes a pulse?
   Missy Heartfeldt

Dear Missy,
When the heart muscle contracts, arteries pulsate or throb as blood surges through them, pushed by the heart. This pulsation is what you can feel by pressing fingers on the wrist or at the side of the neck.

Dear Dr. A. Orta,
My heart beats faster when cute boys are around. Why?
   A Teenager in Peoria

Dear Teen,
An adult normally has a resting heart rate of 70 beats per minute. (This means your heart beats more than 100,000 times a day!) Children have a faster heart beat—about 100 beats per minute. The heart beats even faster than this if you are running, exercising, or falling in love.

Dear Dr. A. Orta,
My heart makes a drumming sound. Is that cool?
   Drummer in Des Moines

Dear Drummer,
The sound of the heartbeat is caused by the slamming shut of heart valves. The first beat is made by the valves between the atria and the ventricles. The second beat is the closing of the valves between the ventricles and the arteries leaving the heart. Don't worry! I'm sure your heart is right on beat!

Reproductive System

Like other mammals, humans reproduce sexually. This means it takes two cells for reproduction to occur: a female egg cell and a male sperm cell. These cells join together in a process called fertilization.

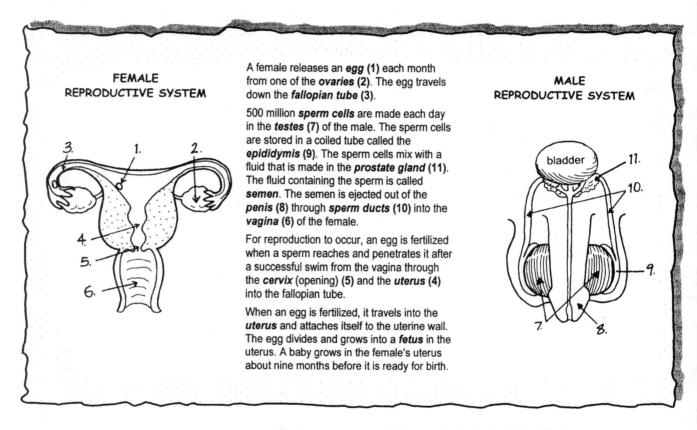

FEMALE REPRODUCTIVE SYSTEM

A female releases an **egg (1)** each month from one of the **ovaries (2)**. The egg travels down the **fallopian tube (3)**.

500 million **sperm cells** are made each day in the **testes (7)** of the male. The sperm cells are stored in a coiled tube called the **epididymis (9)**. The sperm cells mix with a fluid that is made in the **prostate gland (11)**. The fluid containing the sperm is called **semen**. The semen is ejected out of the **penis (8)** through **sperm ducts (10)** into the **vagina (6)** of the female.

For reproduction to occur, an egg is fertilized when a sperm reaches and penetrates it after a successful swim from the vagina through the **cervix (opening) (5)** and the **uterus (4)** into the fallopian tube.

When an egg is fertilized, it travels into the **uterus** and attaches itself to the uterine wall. The egg divides and grows into a **fetus** in the uterus. A baby grows in the female's uterus about nine months before it is ready for birth.

MALE REPRODUCTIVE SYSTEM

bladder

TWINS—How does it happen?

I'm Noel!

I'm Leon!

What causes two babies instead of one?

It can happen two different ways.

Sometimes, after a sperm has fertilized one egg, the egg splits into two at a very early stage of growth. Each part develops into a separate baby, creating *identical twins*. These babies are the same sex and have identical chromosomes.

Sometimes, two eggs are released at the same time, and a separate sperm fertilizes each one. These grow into two separate babies, called fraternal twins. They have different chromosomes, and may be different sexes.

Genetics & Heredity

Dazed and Confused

Dr. Pedia Atrix – Waiting Room

Genetics?
Heredity?
Genes?
Chromosomes?
DNA?
I'm really
confused
about all of it!

Heredity is the passing of traits from parents to their offspring.

A *trait* is a characteristic of an organism (such as height or eye color).

Chromosomes are threadlike structures in the nucleus of every cell that carry the codes for a cell's activity. There are 23 chromosomes in each sperm and each egg. When the sperm and egg combine, they give 46 chromosomes to the baby. The chromosomes are made of proteins and a chemical, *DNA* (deoxyribonucleic acid). Each DNA molecule contains codes that have instructions to control the way the cells work.

Genes are the coded instructions in the DNA. They are the basic units of inheritance. There are hundreds of genes on each chromosome.

Genetics is the study of genes and heredity.

Get Smart Tip # 20

Every person has 23 sets of chromosomes in every cell.

Which sex?

Every person has two sex chromosomes. Males have an X and a Y chromosome. Females have two X chromosomes. All eggs have an X. Half of all sperm have an X and half have a Y. If a sperm with an X chromosome fertilizes an egg, the baby will be a girl. If a sperm with a Y chromosome fertilizes the egg, the baby will be a boy.

What color hair & eyes?

Cells carry two or more genes for each characteristics. Some genes are **dominant**. They tend to overpower the weaker, or **recessive** genes. The gene for dark hair is dominant over the gene for light hair. Brown eyes are dominant over blue eyes. A child born to a dark-haired parent, brown-eyed parent and a blonde, blue-eyed parent is most likely to have dark hair and brown eyes.

175

Diseases & Disorders

There are many things that can go wrong in the body. Henry Hyp O'Chondriac thinks he has them all! Read about these diseases and disorders.

With this *arthritis* my joints are swollen all the time.

I fell off an elephant and have *fractures*—three broken bones in my leg.

I have *cancer*. There is an abnormal division of some cells and they are invading surrounding tissues.

My appendix was so inflamed, it was about to burst. The doctor said I had *appendicitis*.

I have *pneumonia*, an infection in my bronchioles (the smallest tubes in my lungs.)

Hepatitis is my problem. It is a serious infection in my liver.

I have *hemophilia*; my blood doesn't clot properly. It is an inherited condition.

My feet itch from a yucky fungal infection called *athlete's foot*.

I have *gastroenteritis*, an infection of my stomach and intestines.

I have *asbestos poisoning*. It is an environmental disease, caused by breathing toxic asbestos fibers.

I had a *stroke* when a blood clot blocked an artery to my brain.

I need to get treatment for *tetanus*. It's caused by bacteria in an infected wound. The toxins in the bacteria can lead to nerve paralysis.

My congenital disease, *Down's Syndrome*, was caused before birth by an abnormality in my chromosomes.

My *heart attack* happened when arteries to my heart were blocked.

I've lost my voice. My voice box is inflamed. I have *laryngitis*.

My body is unusually sensitive to pollen. This *allergy* is causing my eyes to itch all the time.

I have *influenza*, a nasty ailment caused by a virus.

I have a highly infectious disease, *rabies*. I got it when a raccoon bit me.

I can't breathe well! Due to my *asthma*, my bronchioles are blocked, and I'm struggling for air.

This *pyorrhea* is an infection of my gums.

I have *spinal meningitis*, a serious viral infection of the spinal nerves.

I've got *food poisoning*. Bacteria in something I ate caused it.

I have the *mumps*. My parotid glands are infected.

This *sprain* was caused when I twisted my ankle joint too far.

Defenses Against Disease

The body has a wonderful system of natural defenses to help it fight against diseases and other problems. The body's defenses also help it heal from many ailments.

White blood cells can surround and digest germs.

Antibodies, made by some blood cells, attack and kill germs.

Platelets in blood help the blood clot to stop bleeding.

Saliva kills germs and bacteria in the mouth.

Hairs in the nose filter germs out of the air before they get into the lungs.

Mucous in the nose and throat kill germs before they get further into the body.

Coughing is a reflex that catches and expels harmful substances before they get into the lungs.

Vomiting is a reflex that removes harmful substances from the stomach.

Sneezing is a reflex that keeps dust and germs out of the respiratory tract.

Clean, unbroken skin keeps germs out of the body.

Bone cells are able to multiply and fill in breaks in a bone.

A fever shows the growth of germs. It warns of illness.

Acid in the stomach kills germs.

The liver and kidneys filter toxic substances out of the blood.

Eye-blinking is a reflex that keeps harmful things out of the eyes.

My white blood cells are haywire and my antibodies don't work. I've got acid in my stomach, a fever, coughing, vomiting, and sneezing. I think my kidneys are failing, and my liver doesn't want to live anymore!

The body's defense system can get some outside help from:
- antibiotics to slow or stop bacterial growth
- careful food inspection and preparation
- clean water programs
- vaccines to build up immunity to diseases
- good hygiene such as hand-washing, dental care, and general cleanliness
- surgery to repair damaged organs
- good nutrition to provide the body with substances for growth and repair
- sunshine to provide vitamin D needed by the body for proper use of calcium (to keep strong bones)

How does a vaccination work?
Dead or weakened germs are put into the body. The presence of these germs causes the body to produce antibodies that stay in the body for a long time and protect it from an attack by the live germs. Vaccinations are given by injection or by mouth.

Copyright ©2003 by Incentive Publications, Inc., Nashville, TN.

Health & Fitness

Being healthy means having a body that is working well. Being fit means you are able to use your body to do the things you want to do. There are many aspects to health and fitness. Keeping healthy and fit takes understanding and commitment. The benefits are well worth your time and energy!

Healthy Eating

Your health is definitely affected by what you eat. Make sure you are getting the nutrients you need from these food groups. Don't let fatty, sugary foods or drinks crowd the good things out of your diet.

Every day, get a proper supply of . . .

. . . **protein** in lean meat, fish, seafood, eggs, dairy products, lentils, beans, and nuts. Protein supplies energy. It also helps your body build new cells and repair damaged cells.

. . . **fats** in small amounts in oils, meat and dairy products such as low-fat milk, yogurt, cheeses, and butter. Choose vegetable fats rather than the saturated fats found in animal fats.

. . . **carbohydrates** in grains, breads, rice, cereal, pasta, fruits, and vegetables. Carbohydrates are an important source of energy. Get most of yours from healthy fruits, vegetables, and whole grains.

. . . **vitamins and minerals** from a variety of fruits, vegetables and fresh foods. You need these to keep the systems of your body functioning properly.

. . . **fiber** to keep your digestive system healthy. You can get fiber in fruits and vegetables, whole grains, nuts, seeds, and beans.

. . . **water.** You need about two liters or eight glasses of water every day. If you exercise or work hard and lose water by sweating, drink more!

> The biggest hamburger ever made was 21 feet in diameter and weighed 5,520 pounds.

> Wow! That's a lot of protein!

Doctor! Doctor!

Doctor, doctor! I think I swallowed a roll of film.

Take an aspirin, and we'll see what develops.

Some Aspects of Health & Fitness

✓ a healthy, balanced diet
✓ stamina
✓ flexibility
✓ cleanliness
✓ regular eye, ear and dental care
✓ disease prevention
✓ avoiding dangerous behaviors
✓ good posture
✓ getting plenty of rest
✓ stress management

Other Good Advice

• Go easy on sugar, pastries and pastas with white flour, salt, sweet drinks, and caffeine.
• Avoid fried foods as much as possible.
• Eat food as fresh as possible.
• Maintain healthy weight by moderate, balanced, healthy eating—NOT by crash dieting!

Exercise

Your body needs three different kinds of exercise on a regular basis.

Aerobic Exercise

What is it? Aerobic exercise in which the heart works harder or beats faster for a period of time (20 minutes or longer).

What's the benefit? It builds heart strength and stamina, and improves lung function. Aerobic exercise is also a great way to relieve stress.

What kinds of activities are aerobic? You can get aerobic benefit from any activity that keeps the body (especially large muscles) moving for a period of time: walking, jogging, running, swimming, rope-jumping, cross-country skiing, biking, stair-stepping, etc.

Flexibility Exercise

What is it? Flexibility exercise is movement that keeps your body supple (able to stretch, bend, and twist in different directions).

What's the benefit? It allows the body to perform a wide range of movements without getting injured or sore. It also encourages good posture, improves balance, and reduces stress.

What kinds of activities build flexibility? Any exercise that gently bend and stretch will help increase flexibility. Yoga and other stretching movements are good.

Strengthening Exercise

What is it? Strength is the amount of force produced by muscles. Exercise that strengthens the body builds the amount of force a muscle group has.

What's the benefit? Strengthening exercise builds the size of the muscles and makes them firmer. This gives you more power in those muscles. Strong, firm muscles give you better posture and hold your organs in place well. With more muscle mass, more calories are burned in your body. This helps with weight control.

What kinds of activities build strength? To develop muscles, they need to work hard for a short period of time. Weightlifting, rowing, canoeing, or any other work or exercise that repeatedly uses upper body or lower body muscle strength will build the muscles.

Doctor! Doctor!

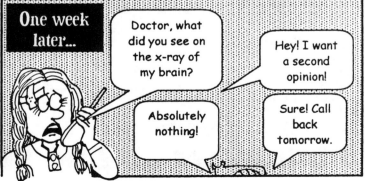

179

First Aid

First aid is immediate and temporary help given to someone who suddenly becomes ill or injured. Knowing first aid can lead you to be more conscious about your own safety. It can also make a critical, perhaps life-saving difference to yourself or someone else in an emergency.

Do This First

1. Quickly examine the victim to check for breathing, bleeding, and a pulse.
2. Keep the victim still. Don't move him or her.
3. Send for help.

Ailment	Symptoms	First Aid
bleeding	blood coming from broken skin	Put direct pressure over the wound with a clean cloth. Get medical help for severe bleeding.
wound	broken skin	Wash with soap and water; cover with sterile bandage.
head wound	dizziness, vomiting, unconsciousness, enlarged pupils, bleeding	Lay victim down with pillow under head. Call for medical help immediately.
shock	pale, cold, clammy skin, irregular pulse, shallow breathing	Keep victim lying down with feet slightly raised. Cover with blankets. Give salt water every 15 minutes if conscious. Get medical help.
fracture	severe pain, swelling	Call for medical help. Apply ice to area. Do not move the victim or injured area.
burns	reddening skin, blistering	For burns that redden the skin, place under cold water. Get medical attention for other burns.
poisoning	stomach sickness, awareness that poison has been drunk	Call a poison control center. Identify the substance and find the container if possible. The container may tell an antidote for the substance.
animal bite	broken skin, possible bleeding	Wash with soap and water. Apply ice to reduce swelling. Check the animal for rabies.
insect sting	itching, swelling	Wash with soap and water. Apply ice to reduce swelling. Scrape the stinger out of the skin gently with a tweezers, knife, or fingernail.
snakebite	pain, swelling, purple color, weakness, nausea, rapid pulse, shortness of breath	Keep victim calm with no movement. Keep bitten area below heart level. Get victim to hospital immediately.
splinter	piece of wood, metal, or glass stuck under the skin	Gently pull the skin away with a sterilized needle. When the splinter is exposed, pull it gently with tweezers to remove the whole splinter. Wash with soap and water.
frostbite	tingling or numbness in nose, feet, hands, or ears; pain; itching; redness	Wrap affected part in a warm blanket or soak in warm (not hot) water. Drink hot fluids.
heat exhaustion	dizziness, headache, skin looks pale and moist	Drink liquids to prevent heat exhaustion. Move victim to a cool spot. Give plenty of water and fruit juice. Give salt water every 15 minutes.
hypothermia	tiredness, shivering, chill, mental confusion, loss of muscle coordination, slurred speech, rigid muscles, loss of consciousness	Treat immediately. Get victim out of wet clothing; cover with blankets; get to a warm place. If victim is conscious, give a warm drink. Get victim to a doctor.
blockage of airway	can't talk, choking, bluish skin, dilated pupils	Try the Heimlich maneuver (abdominal thrust). Call for help immediately.
epileptic seizure	convulsing, jerking	Keep the person from getting hurt. Put something soft under the head.
fainting	pale, moist skin; falling; loss of consciousness	Lay victim flat on the ground and raise the legs. Keep the victim resting for 10 minutes or more.
object in eye	Soreness, swelling, tears	Do not rub the eye. Pull the upper eyelid over the lower eyelid. Wash the eye with an eyewash kit.
nosebleed	sudden bleeding from nose	Keep victim quiet, sitting and leaning forward. Pinch the nostrils together. Apply something cold to nose and face.

──── GET SHARP ──→

on Physical Science

I just love 'mad scientist' movies!

Matter

Matter is the material that makes up all things in the universe. All matter is made up of smaller particles.

An atom is the smallest piece of material that can exist on its own. Every kind of matter is made up of one or more kinds of atoms.

Molecules are combinations of atoms.

Subatomic particles are even smaller than atoms. A *nucleus*, made of *protons* and *neutrons* is at the center. *Electrons* orbit the nucleus.

The different orbits in an atom can hold different numbers of electrons. The first orbit closest to the nucleus holds two. The second orbit holds eight. The third can hold eighteen. However, any orbit that is the last orbit in an atom will only hold eight electrons. Protons carry a positive electric charge. Electrons carry a negative charge. Neutrons carry no charge.

SUBATOMIC PARTICLE SALE

Protons.......... $1.00   Electrons.......... -$1.00
Neutrons.......... No charge

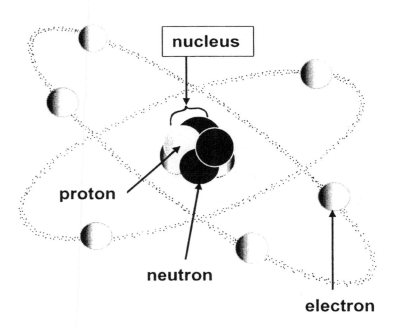

nucleus

proton

neutron

electron

States of Matter

There are three states in which matter exists under normal conditions:

Solids

have a definite size (volume) and shape. The particles are packed together tightly and arranged in a regular pattern.

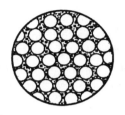

Liquids

have a definite size (volume) but no definite shape. The particles are more active and farther apart than in a solid.

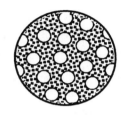

Gases

have no definite size or shape. A gas fills whatever container it occupies. The particles move freely and are far from each other.

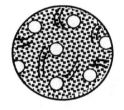

Science Fax

Kinetic Theory

states that all matter is made up of tiny particles in constant motion. The amount of motion and the space between particles are the factors that determine the state of matter. The higher the temperature of the matter, the faster the particles will move.

Changes in States

Freezing is a change from a liquid to a solid state, caused by a lowered temperature. Different substances have different freezing points. Water freezes at 32° F or 0° C.

Melting is a change from a solid to a liquid state, caused by a raised temperature.

Evaporation is a change from a liquid to a gaseous state, caused when the liquid is heated to its boiling point. Different substances have different boiling points. Water boils (and evaporates) at 212° F or 100° C.

Condensation is a change from a gaseous to a liquid state, caused by a lowering of temperature.

Sublimation is a change from a solid to a gas without going through a liquid state.

Oh no! I left the cap off my mother's expensive perfume and it all evaporated away!

Hmmm, evaporation gets blamed for lots of things that people forget to put caps on!

Properties of Matter

Substances have different properties. **Physical properties** are characteristics that can be observed without changing the chemical make-up of the substances. **Chemical properties** are characteristics that describe how a substance will react with another substance.

Physical Properties

Mass is the amount of matter in an object. It is measured in units used to find weight (such as pounds or grams). All matter has mass and takes up space.

When I travel in space, my mass does not change. But my weight changes because the force of gravity is different in space or on different planets.

Weight is the force of gravity pulling on an object. Gravity is expressed in Gs, with gravity on Earth having a value of 1G. The weight of an object is its mass multiplied by the force of gravity (mass x 1G on Earth). On Earth, an object with a mass of 40 pounds will weigh 40 pounds.

Density is the amount of mass packed into a given unit of mass. Density is the mass of the material divided by its volume. It is expressed or measured in cubic units.

Viscosity is the property of a liquid that describes how it pours. Syrup has a greater viscosity than water.

Freezing point and boiling point are the temperatures at which the liquid form of the substance will become solid or the liquid form will become a gas.

Ability to conduct heat (or resist heat) is another property. Substances that conduct heat are called **conductors**. **Insulators** are substances that slow the movement of heat.

Magnetism is the property of attracting certain other substances to itself. Most magnetic substances are metal.

Color, shape, size, shape, odor, and hardness are some other physical properties.

Chemical Properties

A chemical property cannot be observed from looking at the substance. It shows up when the substance interacts with something else. Here are a few examples:

- Charcoal combines with oxygen, during burning, to form carbon dioxide.

- Sodium will combine with chlorine to form sodium chloride, an edible salt. By themselves, each of those substances are not safe to eat!

- When iron combines with oxygen in air, the metal rusts. **Corrosion**, or the ability to rust, is a chemical property. Some substances do not rust. This is a chemical property also.

Fluids & Gases

Fluids are substances that flow. A fluid pushes on objects with a force called a *buoyant force*. This force pushes on ships floating on the ocean and on airplanes flying through the air. The amount of buoyant force is related to the amount of the fluid that is displaced by the object and the weight of the object.

Gases exert force on everything around them. The amount of force a gas exerts per unit of area is called *pressure*. The pressure is caused by busy molecules moving fast inside the container. Every time a molecule bumps against the surface, it exerts a force.

These are some special scientific rules and laws that explain the behavior of fluids and gases.

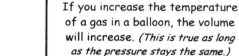

Get Sharp Tip # 22
An object will *sink* if the buoyant force is less that the force of gravity pulling on the object. An object will *float* if the buoyant force is greater than the force of gravity pulling on the object.

Charles' Law
If you increase the temperature of a gas in a balloon, the volume will increase. *(This is true as long as the pressure stays the same.)*

Boyle's Law
If you decrease the amount of gas in a balloon, the pressure will increase. *(This is true as long as you do not change the temperature of the gas.)*

Pascal's Principle
Water pushes up on a floating ship with a force equal to the weight of the ship.

Archimedes' Principle
If you squeeze a plastic bottle of water, the pressure you put on the water will be equal throughout the water. *(There won't be more pressure where you are squeezing!)*

Bernoulli's Principle
In places where air is moving slowly *(at a low velocity)*, the air pressure will be high. In places where air is moving fast *(at a high velocity)*, the pressure will be low.

185

Elements

An *element* is a substance that contains only one kind of atom. An element cannot be broken down by physical or chemical means. There are 103 elements that have been officially named. Most of them occur naturally. Each named element has been given a symbol. There are two main groups of elements: *metals* and *non-metals.* Some elements have characteristics of both and are called *transiton elements*.

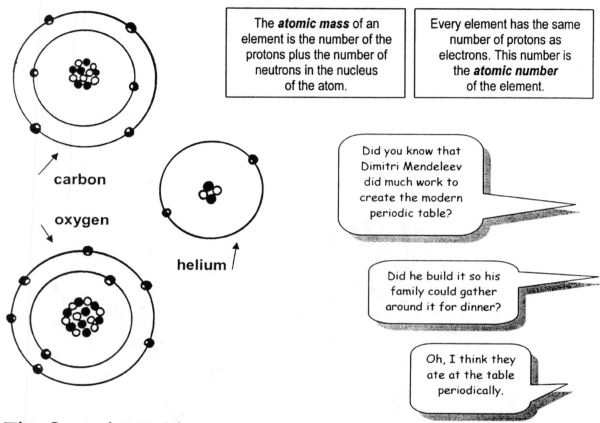

The **atomic mass** of an element is the number of the protons plus the number of neutrons in the nucleus of the atom.

Every element has the same number of protons as electrons. This number is the **atomic number** of the element.

Did you know that Dimitri Mendeleev did much work to create the modern periodic table?

Did he build it so his family could gather around it for dinner?

Oh, I think they ate at the table periodically.

The Periodic Table

The periodic table shows information about the elements. On the table, elements are arranged rows, called *periods* in order of their atomic numbers. They are also arranged in columns, called *groups*. Each group contains elements with similar properties. You can use the periodic table to learn about the elements. Here are a few things the table tells you:

· Argon is a gas.

· Cadmium is a heavier atom than magnesium.

· Gold has 79 electrons.

· Hydrogen has no neutrons.

· The symbol for iron is **Fe**.

· **P** is not the symbol for potassium.

· Aluminum has 14 neutrons.

· Barium has metal properties.

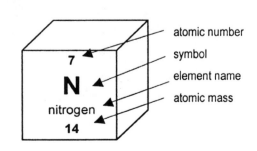

Better Grades & Higher Test Scores / SCIENCE
Copyright ©2003 by Incentive Publications, Inc., Nashville, TN

The Periodic Table

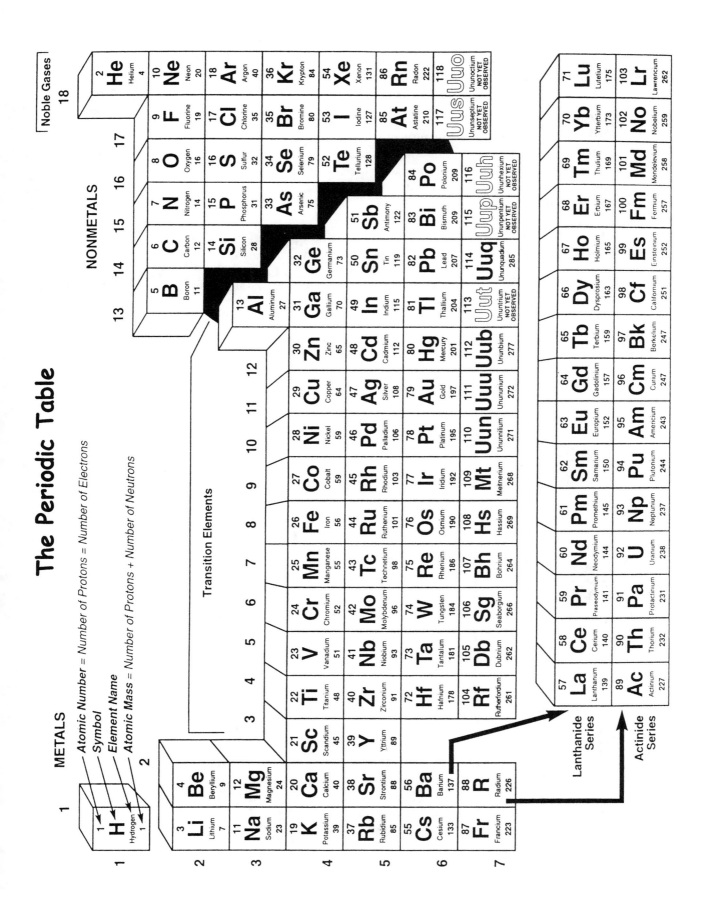

Better Grades & Higher Test Scores / SCIENCE
Copyright © 2003 by Incentive Publications, Inc., Nashville, TN.

Get Sharp: Periodic Table

Compounds

Compounds are formed when two or more elements are joined together chemically. The molecules form a similar element each time that compound is formed. When compounds form, bonds between atoms are rearranged. Atoms of different kinds lose, gain, or share electrons to form bonds between them.

The elements in a compound cannot be separated by breaking, melting, freezing, filtering, evaporating, or any other physical means. When elements join in a compound, each element no longer has the same properties as it did before. The compound has different properties than the original elements.

Each compound has a chemical formula. The formula shows the number of atoms of each element in a molecule of the compound.

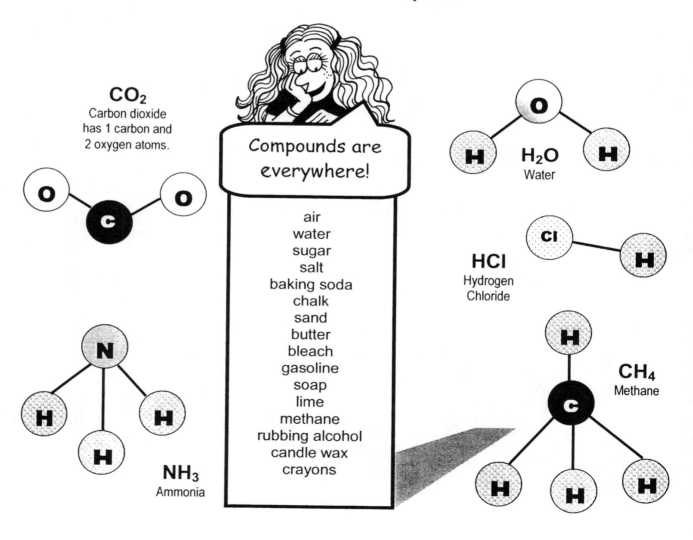

CO_2
Carbon dioxide has 1 carbon and 2 oxygen atoms.

Compounds are everywhere!

air
water
sugar
salt
baking soda
chalk
sand
butter
bleach
gasoline
soap
lime
methane
rubbing alcohol
candle wax
crayons

H_2O
Water

HCl
Hydrogen Chloride

NH_3
Ammonia

CH_4
Methane

Some Common Compounds

Compounds are a part of everyday life. Here are a few of the compounds that get regular use.

Common Name	Formula
marble *(calcium carbonate)*	$CaCO_3$
bleach *(sodium hypochlorite)*	$NaClO$
salt *(sodium chloride)*	$NaCl$
candle wax	CH_2
baking soda *(sodium hydrogen carbonate)*	$NaHCO_3$
sugar *(sucrose)*	$C_{12}H_{22}O_{11}$
methane	CH_4
propane	C_3H_8
chalk *(calcium carbonate)*	$CaCO_3$
ammonia	NH_3
sand *(silicon dioxide)*	SiO_2
gasoline *(octane)*	C_8H_{18}
hydrogen peroxide	H_2O_2
Epsom's salts	$MgSO_47H_2O$
carbon dioxide	CO_2
milk of magnesia *(magnesium hydroxide)*	$Mg(OH)_2$
Freon *(dichlorodifluoromethane)*	CF_2Cl_2
lime *(calcium oxide)*	H_2O
soap *(glycerin)*	$C_3H_8O_3$
copper sulfate	$CuSO_4$
antifreeze *(ethylene glycol)*	$C_2H_6O_2$
water	H_2O

Does the H in H_2O stand for hot water?

If H_2O is hot water, is CO_2 cold water?

I don't know the answer to that question, but I do know that you will be in hot water when Mom finds out that you took your pet alligator into the bathtub!

189

Mixtures

A *mixture* is a combination of 2 or more substances blended together without a chemical reaction. In a mixture . . .

. . . each substance keeps its own properties.

. . . there is no chemical reaction.

. . . there is no repeated chemical make-up or chemical formula.

. . . substances can be separated.

shaving cream

Kinds of Mixtures

A heterogeneous mixture is a mixture in which the particles are not spread evenly throughout. Milkshakes, dirt, and salad dressings with oil, vinegar, and spices are heterogeneous mixtures.

A suspension is a heterogeneous mixture with particles large enough to be seen by the eye or with a microscope. A suspension looks cloudy. The particles in a suspension are not dissolved; they will settle out from the force of gravity. Shaking or stirring will suspend the particles again. The particles can be separated out with filter paper.

Orange juice with pulp is a suspension.

A colloid is a heterogeneous mixture with particles of a size between that of a solution and a suspension. The particles do not settle out with gravity and cannot be filtered out.

Fog, clouds, cheese, jam, and whipped cream are colloids.

A homogeneous mixture is a mixture in which particles of one substance are spread evenly throughout the other substance. Tea, root beer, vinegar, and lemon juice (without pulp) are homogeneous mixtures.

A solution is a homogeneous mixture with very tiny particles of a substance spread evenly throughout. The particles will not settle out when the mixture sits for a while, and a filter cannot separate the particles.

Salt water is a solution.

salt water

Some mystery mixtures have a solution!

cheese

orange juice with pulp

tea

milkshake

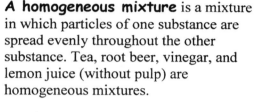

You Can Make Globs of Glorious Green Goop
(and amaze your friends!)

Is it a solid? Is it a liquid? This suspension has characteristics of both! Sometimes it acts like a solid. Sometimes it acts like a liquid.

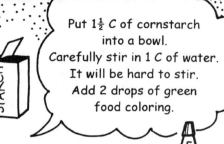

Put 1½ C of cornstarch into a bowl. Carefully stir in 1 C of water. It will be hard to stir. Add 2 drops of green food coloring.

Put some of the goop into your hands. Roll it into a ball and squeeze it until it is hard.

Now, stop squeezing and let it run through your fingers!

Separating Mixtures

Substances combined in a mixture can easily be separated by methods such as

. . . separating metal out of a liquid by using a magnet.

. . . sorting by hand.

. . . shaking smaller particles through a sieve.

. . . settling out solid particles by letting the mixture stand.

. . . passing the substance through filter paper.

. . . distilling to separate two liquids with different boiling points.

Get Sharp Tip # 24

In distillation, one liquid in a mixture evaporates sooner than the other. The vapor of that liquid is collected and cooled back into a liquid.

191

Solutions

A *solution* is a homogeneous mixture in which one substance is dissolved in another. Two or more substances are mixed uniformly.

The solute is the substance being dissolved.

The solvent is the substance in which the solute is dissolved.

An aqueous solution is a solution where water is the solvent. One or more solids, liquids, or gases are dissolved in the water.

(Examples: soda pop and cranberry juice)

A gaseous solution is a solution in which a solid, liquid, or gas is dissolved in a gas.

(Example: air)

A solid solution is a solution in which a solid, liquid, or gas is dissolved in a solid.

(Example: steel, which is carbon dissolved in iron)

FRANNY FRYER'S COOKING SHOW

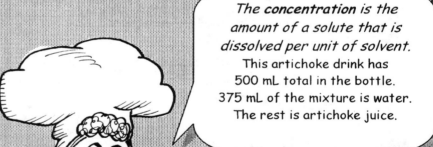

Any of these will cause the solute to dissolve faster:
1. Heat the mixture
2. Crush the solid into smaller pieces.
3. Shake or stir the mixture

Today's Lesson:
How To Make Artichoke Juice

The **concentration** is the amount of a solute that is dissolved per unit of solvent. This artichoke drink has 500 mL total in the bottle. 375 mL of the mixture is water. The rest is artichoke juice.

To find the concentration, divide the amount of the solute by the total amount of the mixture. Then multiply that by 100%.

$$\frac{125\,mL}{500\,mL} = .25 \times 100\% = 25\%$$

Only 25% of this is pure artichoke juice!

192

Solubility

Solubility is the measure of the amount of solute that can be dissolved in a specific amount of solvent at a specific temperature.

An unsaturated solution is a solution that can hold more solute.

A saturated solution is a solution that holds all the solute it can.

A supersaturated solution holds more solute than normal at this temperature. One more drop of solute will crystallize immediately.

Oops! Our artichoke juice is just a tad bitter. Let's add some sugar!

Look at that! The sugar dissolved quite easily. This solution is probably **unsaturated**. Let's be daring and add more sugar.

Four...*five* more tablespoons of sugar. Ah, ha. This new amount just barely dissolves. Our solution is now **saturated** with sugar.

Oops! Another spoonful of sugar won't dissolve at all – it just crystallizes and settles out! The solution is **supersaturated**!

A supersaturated solution, ladies and gentlemen, is one that holds more than it can hold.

Now, that's very interesting. But, why would anyone want to drink artichoke juice?

Acids, Bases, & Salts

Most compounds are one of these: an acid, a base, or a salt.
This is a trio of important kinds of compounds.

> Ions are atoms which have lost or gained some electrons.

Acids

Acids are compounds that produce hydronium ions when they dissolve in water.

An acid. . .

. . . has a strong smell.

. . . has a sour taste.

. . . reacts with metals.

. . . contains hydrogen.

. . . can burn or sting skin.

> That sauerkraut I ate at lunch had a very strong smell and a really sour taste. It sure left a lot of acid in my stomach!

Acidic Substances

vinegar
stomach acid
lemons
oranges
grapefruit
buttermilk
sulfuric acid
bee stings
car batteries

Alkaline Substances

deodorant
antacids
soap
ammonia
drain cleaner
oven cleaner

Bases

Bases are oxides or hydroxides of metal. They produce hydroxide ions in water. Bases that are soluble in water are called **alkali**. If you determine the alkalinity of a substance, you're finding out how strong the base characteristic is.

A base. . .

. . . tastes bitter.　　. . . can dissolve things.

. . . feels slippery.　　. . . can be poisonous.

. . . can burn the skin.

Salts

Salts are products that result when an acid and base **neutralize** each other. The salt is formed when positive ions from a base combine with negative ions from an acid. When the right quantities of an acid and a base are combined, a neutral salt results. The process of neutralization always produces water and a salt.

Testing for Acids and Bases

You can test a solution to find out whether it is an acidic or alkaline (basic) by using an *indicator*, such as litmus paper or phenolphthalein. A good indicator will not only tell if acid or base is present; it will also tell the level of acid or base. Indicators are generally organic compounds that turn different colors in the presence of acids or bases. The indicator's shades of color will vary depending on the strength of the acid or base.

Scientists use a range of numbers, called a pH scale, to tell how acidic or basic a solution is. On the scale, acids range from 1-6 and bases range from 8-14. A solution that tests 7 is neutral. The concentration of the solution (the amount of solute dissolved in the solvent) does not make a difference in the pH reading. A **high pH** shows a high concentration of hydroxide ions. A **low pH** shows a high concentration of hydronium ions.

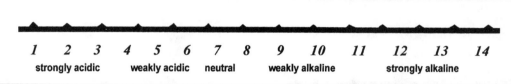

| *1* | *2* | *3* | *4* | *5* | *6* | *7* | *8* | *9* | *10* | *11* | *12* | *13* | *14* |
| strongly acidic | | | weakly acidic | | | neutral | | weakly alkaline | | | strongly alkaline | | |

Indicators

Litmus paper
An acidic solution turns blue litmus paper red.
A basic solution turns red litmus paper blue.
A neutral solution will not change the color of either paper.

Phenolphthalein
An acidic solution turns it colorless.
A basic solution turns it bright pink.

Red Cabbage
Yes! An indicator can be made from the juice of red cabbage. Make your own. *(See page 196.)*

Neutralization

An acid and base can neutralize each other when mixed in just the right quantities. Antacids use a basic solution to neutralize stomach acid that causes indigestion.
A solution or paste of baking soda (base) can relieve the pain of a bee sting (acid).

Ouch!

Help!
I need some
baking soda!

Some pH Values
apple juice	3
orange juice	3.5
milk	6.7
water	7
blood	7.5
sea water	8.5
vinegar	2.8
ammonia	11
stomach juices	1.5
milk of magnesia	10.5

How to Make Your Own Red Cabbage Indicator

You will need:

½ a red cabbage	a large slotted spoon
2 bowls	a sharp knife
a strainer	a grater
water	a clean glass jar with a cover
a small spoon	several small dishes or jars

1. Grate the cabbage into a bowl.

2. Add just enough water to cover the cabbage.

3. Let the cabbage stand in the water. Stir it occasionally.

4. When the water is deep red, use the spoon to remove as much cabbage as you can.

5. Pour the remaining juice through a strainer to get the rest of the cabbage pieces out.

6. Store the juice in a jar with a lid. This juice is your indicator.

Don't drink this – save it!

8. Spoon a few drops of indicator on each sample.

If the sample turns **pink**, the substance is an *acid*.

If the substance turns **green or blue**, the substance is a *base*.

(You may have to wait up to 2 hours for the color to change.)

7. Put small samples of substances in little dishes or jars.

Some things to test:

lemon juice	antacid liquid or tablet	tomato
baking soda	laundry soap	liquid from canned fruits
baking powder	soda pop	cooking water from boiled vegetables
vinegar	milk of magnesia	cottage cheese
orange juice	egg white	tea
buttermilk	cranberry juice	coffee

Changes in Matter

There are many ways that matter changes. The changes all fall into two categories. A *physical change* is a process that does not change the chemical make-up in a substance. After the change, the material still has the same properties. Only the size, shape, color, or state has changed. A *chemical change* does involve a change in the chemical make-up of the substance. After the change, it no longer has the same properties. It has become a new substance. In any change, physical or chemical, some kind of energy must be applied to the substance in order for the change to occur.

Physical Changes

breaking glass

separating rocks from sand

evaporating water

freezing popsicles

rain condensing from clouds

clothes drying in a dryer

blowing the top off a dandelion

mixing up a milkshake

slicing bread

making chocolate milk

whipping cream

making lemonade

blowing a glass sculpture

Chemical Changes

baking a cake

burning a candle

striking a match

frying an egg

garbage rotting

a bicycle rusting

an old log decomposing

plants making oxygen

toast getting crisp and brown

food digesting

bread molding

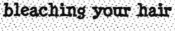
bleaching your hair

197

Motion & Force

If you want to ride a skateboard, relocate a tent, lift weights, pull up your socks, open a jar, fly a kite, or wiggle your ears, you need a force. Nothing moves until a force acts on it. Nothing stops moving until a force acts on it.

Motion is a change in position of some object or substance.

A force is a push or pull acting on something. A force is needed to start a motion, stop a motion, make something go faster or slower, or change directions of object or body. Some forces act directly to push or pull on object. Other forces, such as gravity and magnetism, act at a distance.

Get Sharp Tip # 26
The unit used for measuring force is the newton (N). A force of 9.8 N pulls upon each kilogram of mass in an object at rest on Earth.

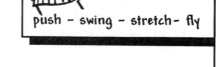
push – swing – stretch– fly

Force makes it possible to:
push   lower   throw   bend   pull
lift   mix   squeeze
stretch   raise   twist   press

Gravity is the basic force in the universe. Every body in the solar system has a force that pulls things to itself. That's gravity! On Earth, it keeps people and objects from flying off into space. An object's weight depends on the force of gravity. The pull of gravity is different on different bodies in space, so weight varies on different planets or moons.

kick – twist– soar – pump

the awesome pull of gravity

Centrifugal force and centripetal force are two forces acting on the object that spins around a second object. Centrifugal force pushes the object outward. Centripetal force pulls the object inward.

Remember:
Don't let gravity get you down!

Net Force

When the forces acting on an object are unbalanced, a **net force** results. One force is stronger than the other. A net force is needed to change the motion of an object (to change its speed or direction). During a tug of war, there must be a net force for one team to pull some of the other team members across the line.

Balanced Forces

When something is not moving, the forces on it are the same. Two forces are acting on the object equally from opposite directions. If something isn't moving, the forces are always **balanced**. During a tug of war, the forces are balanced when neither team is able to pull any of the other team members across the line.

Science Fax

Pressure is a force that presses evenly on a surface. Air exerts pressure all over Earth's surface. Its force is equal to 1 kilogram per square centimeter.

May the Force Be With You!

Zelda lies down on a puffy mattress.

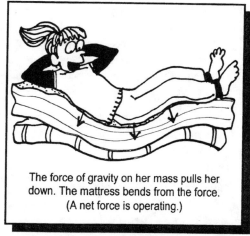

The force of gravity on her mass pulls her down. The mattress bends from the force. (A net force is operating.)

May the force be with you, Zelda!

The force of the mattress pushes up against her. Zelda stops sinking into the mattress when the force of the mattress pushing equals the force of gravity pulling down on her. (The forces are balanced.)

More Things to Know About Motion & Force

Friction is the force that opposes motion between two touching surfaces. *A sled gets to the bottom of a hill of slippery snow. It keeps moving ahead for a while, but friction causes it to stop eventually.*

Speed is the rate of motion of a body. It is expressed in distance per units of time. *A fast train might travel at a speed of 100 miles per hour.*

Velocity is speed in a particular direction. *A horse gallops around a circle at 10 miles per hour. His speed is pretty fast, but his velocity is zero because he ends up where he started. The horse does not make progress in any direction.*

Some Facts About Bob's Sled

Inertia keeps the bobsled sitting still until someone pushes it. The harder it's pushed, the greater the velocity will be.

The sled accelerates as it heads downhill. It picks up more momentum with each second.

The bullet shape reduces the air resistance against the sled. The sled's runners melt the ice and reduce the friction of the snow against the sled.

Bob, it's time to go home!

I'll be there in a momentum!

Acceleration is the rate of change in velocity when the velocity increases.
Deceleration is the rate of change in velocity when the velocity decreases.
A car accelerates from 20 mph to 40 mph. It decelerates from 40 mph to 0 mph.

Inertia is the property of a body to resist any change in velocity. *A child pulls a sled along behind her on the snow. When she stops, the sled keeps moving and hits the back of her legs.*

Momentum is the quantity of motion in a moving body. It is found by multiplying the mass and velocity. *A soccer ball kicked by an adult will have greater momentum than a ball rolled by a child; even though the balls have the same mass, the velocity is greater.*

Resistance is any opposition that slows something down or prevents movement. It is a form of friction. *When you ride your bike into a strong wind, the wind resistance slows your velocity.*

Laws of Motion

In 1687, an English scientist named Sir Isaac Newton published three rules that describe the ways force changes motion. These are known as *Newton's Laws of Motion*.

Get Sharp: Motion & Force

Energy

Energy is everywhere. We can feel it as heat, see it as light, and hear it as sound. We use energy that is caused by movement of wind and water. We use energy produced by electricity or by chemical reactions inside and outside our bodies. **Energy** is the ability of matter to do work or to make a change in itself or its environment. Every object has energy because of its position or motion. Energy has many forms, and it can change from one form to another.

Kinetic energy is the energy of motion. An object's kinetic energy depends on its mass and velocity. Objects with greater velocity or mass have more kinetic energy. *A moving bus or jumping frog has kinetic energy.*

Potential energy is stored energy. Every resting object has potential energy. This means that if the object were sent into motion it would have kinetic energy. *A roller coaster stopped at the top of a steep hill has potential energy.*

Mechanical energy is potential and kinetic energy combined in lifting, stretching, or bending. *You can see mechanical energy if you watch kids pulling taffy.*

Work is the transfer of energy that results from motion. *Picking up a rock is work.*

Radiant energy is energy that can travel through space in the form of waves. *The Sun gives off radiant energy.*

Chemical energy is energy released during a chemical reaction. *The digestion of food is a chemical process that releases energy for body activity. Burning of wood is a chemical reaction that releases energy as heat and light.*

Thermal energy is the total energy in all the particles of an object. The amount of energy depends on the temperature of the object. *Boiling soup has more thermal energy than an ice cube.*

Solar energy is energy that is trapped from the Sun. The energy can be stored and turned into electricity.

Nuclear energy is energy stored in the nucleus of every atom. This energy is extremely powerful. Most nuclear energy produced comes from the atoms of certain elements such as uranium. *This energy is released by **fission** (splitting of the atoms) or **fusion** (joining together nuclei of atoms).*

What do British nuclear physicists eat for lunch?

Fission chips, of course!

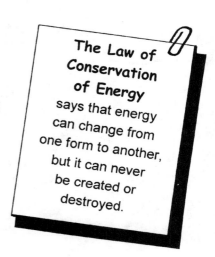

The Law of Conservation of Energy says that energy can change from one form to another, but it can never be created or destroyed.

Energy Fax

- The Sun is the most powerful source of energy in our solar system. It has provided heat and life for billions of years. Its energy is released by the burning of about 20 billions tons of hydrogen every minute.

- All living things depend on the Sun for their energy.

- All energy becomes heat. Rub your hands together and feel the heat that results from the energy you're using to move your hands together.

- Energy can change forms, but it cannot be lost.

- Energy can be transferred from one material to another. Food is warmed when heat energy from the oven is transferred into the food.

- Some energy is lost in every transfer. A light bulb uses energy to give light, but most of the energy it uses is lost as heat.

- Temperature is the measure of the average kinetic energy in the particles of an object.

- Specific heat is the amount of energy needed to raise the temperature of one kilogram of a material one degree Celsius. Different materials have different specific heats. The specific heat of water is 4190° C. The specific heat of copper is 380° C.

- Much of the energy used in the world comes from the burning of fuels such as coal, oil, natural gas, or gasoline.

- Energy is often used to generate electricity. The power of wind, or power from burning fuels is changed into heat and light energy.

- Energy is measured in joules (J). One joule is the amount of energy needed to move a 2-kilogram weight one meter in a second.

THE OBEDIENT CAN

You can make a can roll away from you and return when you call it!

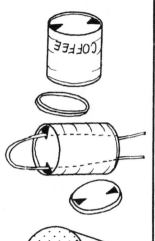

1. Get a coffee can with a plastic lid. Use a can opener to punch a hole on each side of the end of the can.

2. Make one cut in a long, sturdy rubber band. Follow the diagram to thread it through the holes in the can.

3. Attach a heavy bolt to the center of the rubber band. Tie it with string.

4. Punch two holes in the plastic lid to match the holes you punched in the can. Thread the rubber band through those holes and tie them securely on the outside of the lid.

5. Roll the can away from you. When the can stops or slows down, command the can, "Come back!" The can will return to you.

How it Works

When the can rolls away, the kinetic energy causes the rubber band to twist. The twisted band stores potential energy. When the can stops, it is ready to turn that potential energy back into energy of motion (kinetic energy).

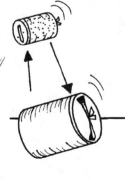

COME BACK!

203

Heat

Heat is energy that is transferred from an object at a higher temperature to one at a lower temperature. The energy of heat is called *thermal energy*. Heat is measured in degrees. There are two commonly used scales: Celsius and Fahrenheit. On the Celsius scale, water freezes at 0° C and boils at 100° C. On the Fahrenheit Scale, water freezes at 32° F and boils at 212° F.

Transfer of Heat

Heat moves from one object or substance to another in one of three ways:

Conduction is transfer of heat from particle to particle when two substances come in contact with each other. Particles at high temperatures vibrate faster and further. They collide with the cooler particles and share their energy. Then the slower, cooler particles start to move faster.

When you put ice cubes into a drink, heat from the warmer drink transfers to the cubes. When you put a pan of soup on a hot stove, heat from the stove transfers to the cooler soup.

A *conductor* is a substance that transfers heat. Some materials conduct heat better than others. Some good conductors are copper, iron, silver, and water. Plastic, gas, glass, and wood are poor conductors.

An *insulator* is a substance that reduces the transfer of heat. Poor conductors are often used as insulators.

Convection is transfer of heat by movement of matter. Convection is the way heat travels in fluids (liquids and gases). When a fluid is heated, its density and weight decreases. Cooler, denser fluid sinks, pushing the warmer, less dense fluid upwards. This motion of the fluid creates convection currents that spread heat throughout the substance.

Radiation is the transfer of heat that does not need matter. The heat from the Sun travels across millions of miles of empty space by radiation. Heat can transfer by radiation through air, as well. When you hold your hands out to warm them near a hot campfire (without touching the fire), you are enjoying heat transferred by radiation.

Work & Machines

There's lots of work to be done in the world. It's a good thing we have machines, because much of the work is hard. Machines make all kinds of work easier.

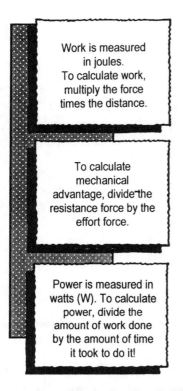

Work is measured in joules. To calculate work, multiply the force times the distance.

To calculate mechanical advantage, divide the resistance force by the effort force.

Power is measured in watts (W). To calculate power, divide the amount of work done by the amount of time it took to do it!

Work is the transfer of energy as a result of motion.

A machine changes the amount or direction of the force that must be exerted to do work. The reason for a machine is to convert a small amount of force into a much larger force.

A simple machine is a machine that consists of only one part.

A compound machine is a combination of two or more simple machines. Many of the machines in the world combine several simple machines.

Effort force is the amount of force that must be applied to the machine.

Resistance force is the force exerted by the machine.

Mechanical advantage is the number of times a machine multiplies the effort force.

Power is the rate of doing work. It is the amount of work done per unit of time.

The Six Simple Machines

LEVER
The lever is a bar that is free to pivot about a fixed point called a fulcrum. A lever changes the amount and direction of the effort force.

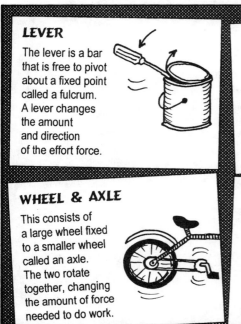

PULLEY
A pulley is an arrangement of one or more wheels and rope. A pulley changes the amount of force and/or the direction of force needed to do work.

INCLINED PLANE
The inclined plane is a slanted surface used to raise objects. A pulley changes the amount of force needed to do work.

WHEEL & AXLE
This consists of a large wheel fixed to a smaller wheel called an axle. The two rotate together, changing the amount of force needed to do work.

SCREW
The screw is an inclined plane wound around a cylinder. It has a spiral thread along the edge. The screw changes the amount of force needed to do work.

WEDGE
The wedge is an inclined plane with at least one sloping side. The wedge changes the amount of force needed to do work.

Better Grades & Higher Test Scores / SCIENCE
Copyright ©2003 by Incentive Publications, Inc., Nashville, TN.

Electricity

A phone rings. A flashlight beam helps you find your way to your tent. A siren warns of an approaching ambulance. A CD player booms out your favorite tune. The click of a mouse connects you to an awesome website. A flat screen on the grocery counter reads the price of your potato chips. A startling streak of light splits the dark sky during a thunderstorm. All of these are amazing tricks and displays of electricity.

Electricity is the energy resulting from the flow of electrons. Electricity is all around you every day, showing up naturally in spectacular displays and powering all sorts of conveniences.

An electric charge results when an object has too many or too few electrons. Every atom is made of particles that have electric charges. Electrons have a negative charge and protons have a positive charge. Electrons move through matter. When an object gains electrons it becomes negatively charged. An object that loses electrons becomes positively charged.

Static electricity is an electric charge built up in one place. Clothing and other objects can build up static electricity by rubbing against each other. You've seen this when you rub a ballon against clothing and place it near your head. Your hair stands on end!

A conductor is something that allows electricity to flow through it easily.

An insulator is a substance that is a poor conductor of electricity.

IF LIFE GIVES YOU LEMONS,

Make Your Own Lemon Battery

You need: light bulb   copper wire   2 lemons   a knife   steel wire or paper clips

1. Squeeze and roll the lemons to get them juicy. Do this gently; do not break the skin.

2. Cut a 10-inch strip of each wire. Remove any insulation from the ends of the wires so they are bare. Cut a 3-inch strip of each wire. Twist these together.

3. Attach one end of each wire to the light bulb holder. Place the other end of each wire deep into one of the lemons. Place the small copper-steel wire combination into the lemons as shown.

How it works!

The chemical reactions between the lemon juice and the two metals pushes electrons to flow through the circuit you've made. If the lemon battery is not strong enough to light the bulb, touch the two wires (from the bulb) to your tongue to complete the circuit. You'll feel the tingle of the electricity produced by the lemon battery.

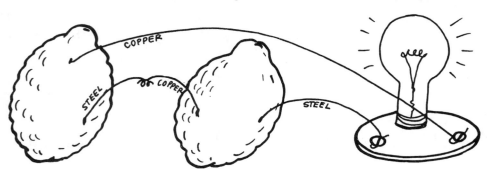

Better Grades & Higher Test Scores / SCIENCE
Copyright ©2003 by Incentive Publications, Inc., Nashville, TN.

Electric current is a steady flow of electrons or other charged particles through a conductor. For practical use, such as to run a machine or light a bulb, current needs to flow continuously. Some source of force, such as a battery, is needed to keep the electrons moving.

Direct current (DC) is the flow of electrons in one direction. When the force pushing the electrons is a battery, they flow from a negative terminal and flow to a positive terminal.

Alternating current (AC) is current that changes directions. In North America, power companies distribute alternating current that changes directions 120 times a second. This is called 120 current.

Potential difference is the difference in potential energy between the electrons in one place and the potential energy in the electrons in another place. *Voltage* is another word for potential difference.

Lightning is the most spectacular example of natural electricity. A flash of lightning can produce 100 million volts of electricity.

I got a shock when I touched the metal doorknob, but I got no shock when I touched the window. So there you have it! That means metal is the greatest conductor!

I thought Leonard Bernstein was the greatest conductor.

Resistance is a measure of how hard it is to push electrons through a conductor. The resistance of a conductor depends on the thickness, length, and type of material of the conductor. Some materials, such as wood, glass, and plastic, have high resistance to the flow of electricity. Others, such as copper and water, have little resistance.

Electric power is the rate at which a device (such as a machine) changes electricity into another form of energy.

Electrical energy is the energy used by an electrical device. The amount of energy used by an electrical device depends on the amount of power delivered and the length of time the power is used.

Lightning is the release of electricity that has been built up inside of clouds. The electricity finds a path through drops of water and objects such as a tree or building.

Electric Conversations

Electric Circuits

A *circuit* is an unbroken path formed by electrical conductors. In order to flow, a current must have an uninterrupted loop of electrical conductors. The electricity will flow from negative to positive terminals in these batteries if the wires are connected correctly. If a circuit is broken at any point, the electricity will stop flowing.

A series circuit has only one path for the current to follow. The current is the same in every part of the circuit. If any part of the circuit is broken, the current stops.

A parallel circuit has two or more separate branches for current to flow. If the circuit is broken in one branch, the current will still flow to other parts of the circuit.

A switch is a device on a circuit that can open and close the pathway to start and stop the flow of electricity.

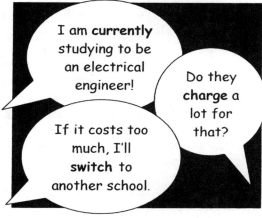

Get Sharp Tip # 28
Electrons must be able to flow back to the battery or other source of force pushing them along.

I am **currently** studying to be an electrical engineer!

Do they **charge** a lot for that?

If it costs too much, I'll **switch** to another school.

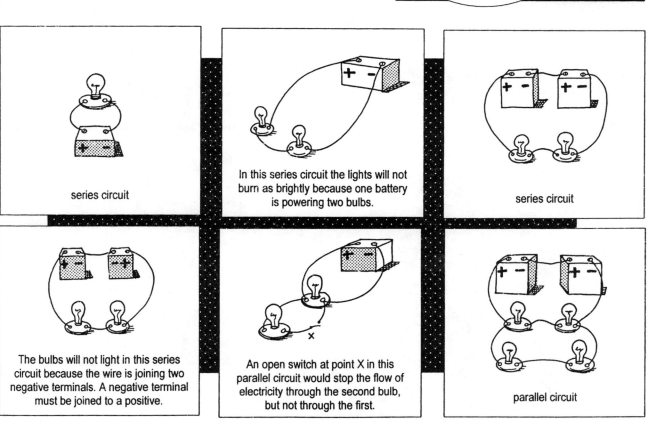

series circuit

In this series circuit the lights will not burn as brightly because one battery is powering two bulbs.

series circuit

The bulbs will not light in this series circuit because the wire is joining two negative terminals. A negative terminal must be joined to a positive.

An open switch at point X in this parallel circuit would stop the flow of electricity through the second bulb, but not through the first.

parallel circuit

Get Sharp: Electricity

Magnets

A *magnet* is an object that attracts other magnetic materials. The Earth is a giant natural magnet. Lodestone (magnetite), one of Earth's materials, is a natural magnet. Some other materials can be made into permanent magnets by exposing them to other magnets. These substances, such as iron, aluminum, cobalt, and nickel, can hold their magnetism for a long time. Also, some substances can be made into magnets by running electricity through them.

Magnetic poles are the opposite ends of a magnet. The magnetic force is strongest at these locations. The poles are labeled *north* and *south*. When a magnet is held near a magnetic substance such as iron filings, most of the filings will cling to the magnet at its poles.

> **Unlike poles** (north and south) will attract each other.
> **Like poles** (north and north or south and south) will repel one another.

The magnetic field is a region around the magnet where the magnetic force acts. This region is near the poles, but extends out from the actual magnet.

Magnetic Needles

Get a large needle. Hold it carefully in one hand. Stroke it against a magnet 20 times in one direction. Only use one end of the magnet; don't switch ends when stroking!

Now, use the needle to pick up some straight pins or paper clips.

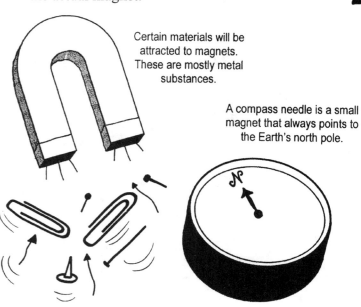

Certain materials will be attracted to magnets. These are mostly metal substances.

A compass needle is a small magnet that always points to the Earth's north pole.

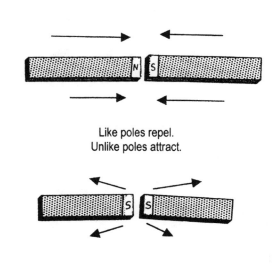

Like poles repel.
Unlike poles attract.

Better Grades & Higher Test Scores / SCIENCE

Electromagnets

An *electromagnet* is a coil of wire with electric current flowing through it. Electromagnets have wide use in the world of electronics and other technology. They are used for heavy machinery and motors, and can be found working wonders in cars, trucks, trains, televisions, and tape recorders.

In an electromagnet, the ends of the coils are the *poles*. Often an electromagnet has an iron *core* in the center of the coil. This makes the magnet stronger.

Get Sharp Tip # 29

An electromagnet becomes stronger with more turns of the wire in the coil or with a larger electric current.

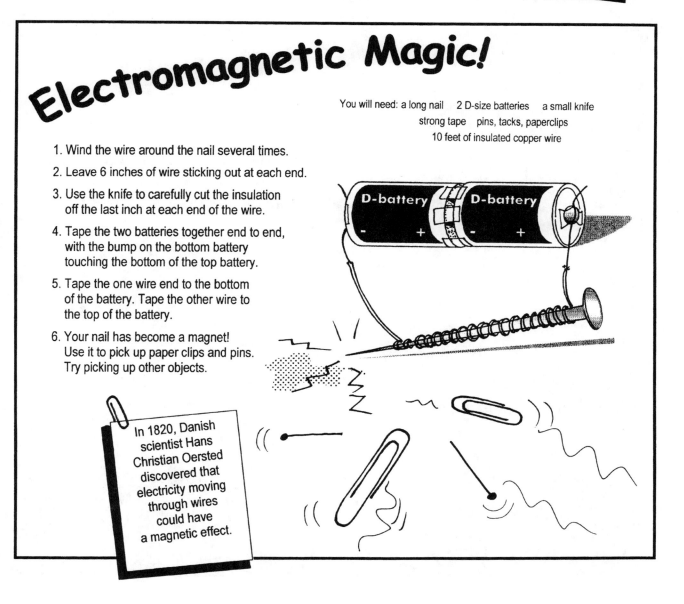

Electromagnetic Magic!

You will need: a long nail   2 D-size batteries   a small knife
strong tape   pins, tacks, paperclips
10 feet of insulated copper wire

1. Wind the wire around the nail several times.

2. Leave 6 inches of wire sticking out at each end.

3. Use the knife to carefully cut the insulation off the last inch at each end of the wire.

4. Tape the two batteries together end to end, with the bump on the bottom battery touching the bottom of the top battery.

5. Tape the one wire end to the bottom of the battery. Tape the other wire to the top of the battery.

6. Your nail has become a magnet! Use it to pick up paper clips and pins. Try picking up other objects.

In 1820, Danish scientist Hans Christian Oersted discovered that electricity moving through wires could have a magnetic effect.

Waves

When you hear the word *waves*, you probably think of the rolling surf against a beach, whitecaps on a lake, or the lone surfer carried along on a giant wall of water. The waves that rise from water are only one kind of wave; there are many others. Light rays, radio waves, microwaves, X-rays, sound waves, and cosmic rays are a few of the other kinds of waves.

A *wave* is a rhythmic disturbance that carries energy. All waves transfer energy from one place to another. As it moves, the energy from a wave affects everything in the path of the wave. In a *transverse wave*, matter moves at right angles to the direction the wave is traveling. In a *compressional wave*, matter moves in the same direction as the wave travels.

Wave Characteristics

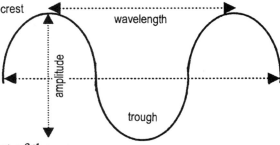

Waves with a long wavelength have low frequencies. Waves with short wavelengths have high frequencies.

The greater the energy a wave has, the greater its amplitude will be.

The crest is the highest point of the wave.

The trough is the lowest point of the wave.

The wavelength is the distance between one point on a wave the same point on the wave next to it.

The amplitude is the greatest distance particles in the wave rise or fall from their resting position.

Frequency is the number of waves that pass a given point in one second. Frequency is measured with a unit called a *hertz* (Hz). One hertz equals one wave passing a point per second.

Velocity is the distance traveled by any point on the wave in one second. *(Velocity equals wavelength multiplied by frequency.)*

Vibration is the up and down movement of the wave.

What do physicists do at football games?   They do the wave!

Better Grades & Higher Test Scores / SCIENC
Copyright ©2003 by Incentive Publications, Inc., Nashville, TN

The Electromagnetic Spectrum

Electromagnetic waves are transverse waves that travel at the speed of light in a vacuum. As these waves travel, they transfer energy. The transfer of energy by electromagnetic waves is called **radiation**.

The **electromagnetic spectrum** is an arrangement of the different waves according to their lengths. The only visible parts of the spectrum are light waves (rays).

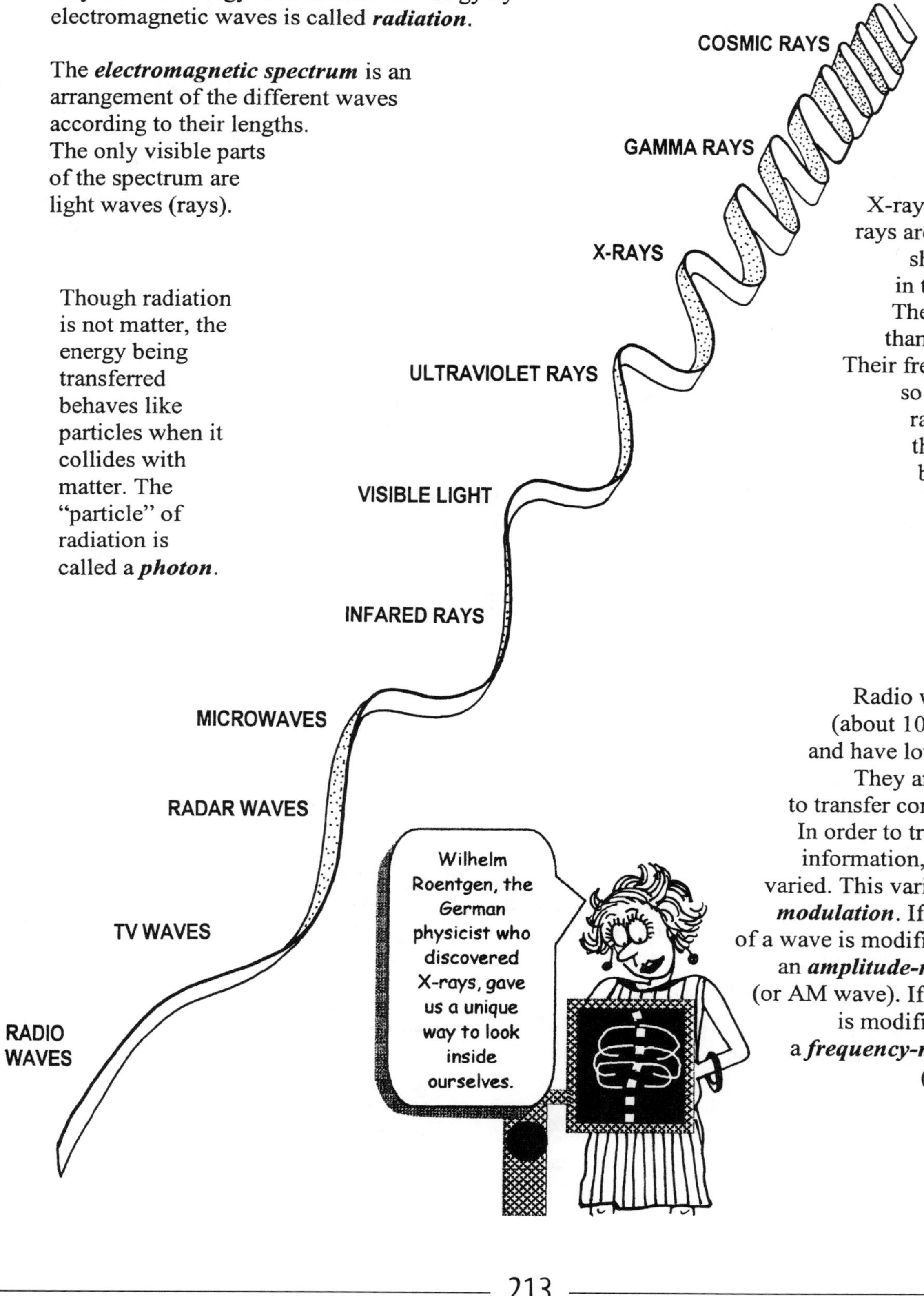

Though radiation is not matter, the energy being transferred behaves like particles when it collides with matter. The "particle" of radiation is called a **photon**.

COSMIC RAYS

GAMMA RAYS

X-RAYS

ULTRAVIOLET RAYS

VISIBLE LIGHT

INFARED RAYS

MICROWAVES

RADAR WAVES

TV WAVES

RADIO WAVES

X-rays and gamma rays are some of the shortest waves in the spectrum. They are shorter than 0.0001 mm. Their frequencies are so high that the radiant energy they carry can be harmful to human cells.

Radio waves are long (about 10 cm or longer) and have low frequencies. They are mostly used to transfer communications. In order to transmit a lot of information, the waves are varied. This variation is called **modulation**. If the amplitude of a wave is modified, it is called an **amplitude-modified wave** (or AM wave). If the frequency is modified, it is called a **frequency-modified wave** (or FM wave).

Wilhelm Roentgen, the German physicist who discovered X-rays, gave us a unique way to look inside ourselves.

213

Light

Light is the part of the electromagnetic spectrum that is visible. Our eyes can see a section of the spectrum containing a range of wavelengths that appear as different colors. White light is a mixture of all these colors. We see objects because they reflect light.

A **transparent** object is made of material that allows the light to pass through, and you can see through it.
A **translucent** object is made of material that you cannot see through, but light can pass through it.
An **opaque** object absorbs light. Light does not pass through it, and you cannot see through it.

Reflection

When light waves are not absorbed by an object or do not pass through it, they may bounce off the object. This *bouncing off* action is called **reflection**.

Light is reflected at an angle. The waves that strike the object are **incident waves**. The angle at which they strike is the **angle of incidence**.

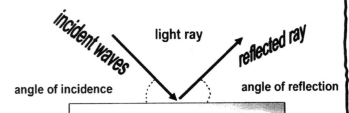

light ray

angle of incidence   angle of reflection

The waves that bounce off are **reflected waves**. The angle at which they bounce off is the **angle of reflection**. These two angles are always equal.

Refraction

Refraction is the bending of light waves. When light waves (rays) pass from one substance into another at any angle other than a right angle, they are bent. The bending distorts the image that is seen by the eye.

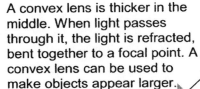

Lenses

A *lens* is a curved, transparent object. When light passes through a lens, the direction of the light is changed. Often the lenses produce an image that appears smaller or larger than the original object.

A concave lens is thinner in the middle. When light passes through it, the rays are bent away from each other. A concave lens can make objects appear smaller.

A convex lens is thicker in the middle. When light passes through it, the light is refracted, bent together to a focal point. A convex lens can be used to make objects appear larger.

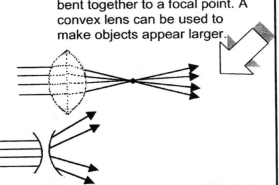

Better Grades & Higher Test Scores / SCIENCE

How To Spy Around Corners

Why does it seem as if granny can see around corners?

Because she can!

Ah ha! Sneaky Sam has his fingers in the cookie jar again!

A periscope allows grandma to use the properties of light to see around corners. Follow the directions to make a periscope of your own.

milk carton

Mirrors at 45° angles

You will need:
a quart-sized milk carton   a scissors
tape   2 identical pocket mirrors

1. Open the top of the milk carton. Tape two mirrors at a 45° angle, parallel to each other, as shown.
2. The top mirror should be faced down. The bottom mirror should be faced up.
3. Cut a hole near the bottom, facing one mirror, as shown.
4. Cut a similar hole at the same location near the top, as shown.
5. Make sure that both mirrors and holes are in the same locations in relation to the top or bottom of the carton, as shown.
6. Use your periscope to see places without being seen.

Color

Different waves within the range of visible light have different lengths. These waves reflect with different colors. A rainbow shows all the colors of the visible spectrum when drops of rain in the air refract sunlight. This is one time when we can see the red, orange, yellow, green, blue, indigo, and violet rays that combine to make white light. Another way to see the colors of the light is to bend the light with a prism. We see individual colors in objects because the object has reflected the waves that have the length that shows that color.

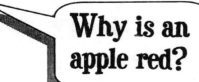

Why is an apple red?

The molecules in the skin of the apple absorb all the waves in light except the red ones. These rays are reflected back to our eyes – and the apple looks red!

215

Sound

The beat of a drum, a howling wind, a screeching owl, a whining child, a loud rock concert, a ringing phone, a sonic boom, a dripping faucet—these sounds and all others are made by the vibration of some object or objects. When something vibrates, it moves back and forth. The molecules in the object vibrate and cause the air around the object to vibrate, too. This sets up a sound wave that travels through the air in a series of ripples moving out from the object in all directions. Sound waves are **compressional waves**, waves whose particles move in the same direction that the wave travels.

Sound cannot travel through a vacuum. This means that there is no sound in space. Sound must have matter for its transmission. The velocity of sound (the distance it travels in a given time) depends on the kind of matter through which it is traveling. Sound travels faster through liquids and solids than through gases.

The Doppler Effect

When a sound-making object or the listener is moving, there is a change in the frequency of the sound waves. This change in wave frequency is called the **Doppler effect**. It affects the pitch of the sound (how high or low the sound is). The motion of a plane towards you crowds the sound waves together, increasing the frequency and causing the wavelengths to shorten. Shorter wavelengths cause the higher pitch you hear. When the jet moves past you, the waves are farther apart, the frequency of the waves is decreased, and the pitch appears lower.

The Doppler effect also occurs when the listener is moving. If you walk past a lawn mower, you'll find that the pitch is higher as you approach. The sound waves are striking your ears more frequently and wavelengths are shorter. As you pass the lawnmower, the pitch will be lower because the waves will strike your ears less frequently.

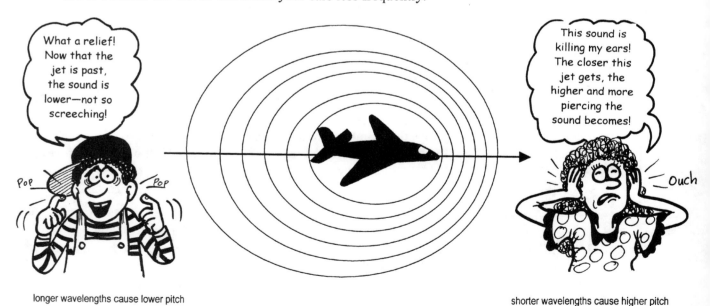

longer wavelengths cause lower pitch

shorter wavelengths cause higher pitch

The Language of Sound

The **amplitude** of a sound is the greatest distance the particles in a wave rise or fall from their rest position. A sound wave with large amplitude will carry a loud sound.

The **intensity** (loudness) of a sound depends on the amplitude of the sound waves. The bigger the amplitude, the more intense the sound will be. The loudness of a sound can differ for different people. Loudness describes the way a person responds to a sound's intensity. A sound that seems loud to one person may not be loud enough for someone else.

A **decibel (dB)** is the unit used to measure loudness or intensity of a sound. Normal talking has a loudness of about 30 dB. The quietest sound that can be heard is 0 dB. Sounds of 120 dB or above cause pain to the human ear.

The **frequency** of a sound is the number of waves that pass a certain point in a second. Frequency is measured in *hertzes* (Hz). Humans can hear sounds ranging from 20 and 20,000 Hz. Some animals can hear sounds with frequencies higher than those we can hear. As the frequency of a sound wave increases, the wavelength decreases.

Pitch is the highness or lowness of a sound. The pitch of a sound is related to its frequency. Sounds with high pitches have high frequencies and sounds with low pitches have low frequencies.

The **velocity** of a sound wave is the speed and direction the sound moves. Velocity depends on the kind of matter the sound is traveling through. Sound travels faster through liquids and solids than it does through gases. Sound travels at about 332 meters per second through dry air at 0° C. The velocity increases as the temperature increases. Sound travels faster through warm air than through cold air.

Tone quality has to do with the differences among sounds that have the same pitch and loudness. Tone qualities differ depending on the source of the sound. The tone quality of a note played on a piano is different from the tone of the same note played on a tuba.

Noise is sound that has no definite pitch or regular wave pattern. Noise can also be defined as unpleasant sound. Different people have different ideas about what sounds are pleasing.

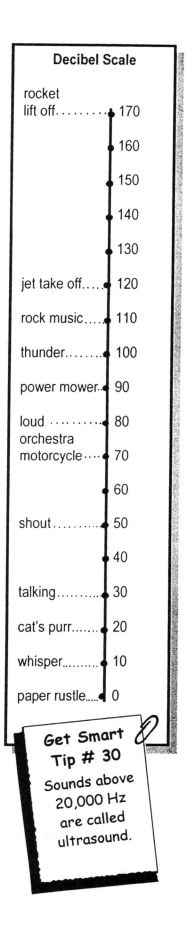

Decibel Scale

rocket lift off......... 170
160
150
140
130
jet take off..... 120
rock music..... 110
thunder........ 100
power mower.. 90
loud 80
orchestra
motorcycle.... 70
60
shout......... 50
40
talking......... 30
cat's purr....... 20
whisper......... 10
paper rustle..... 0

Get Smart Tip # 30
Sounds above 20,000 Hz are called ultrasound.

217

Fooling Around with Sound

Make your own stringed instrument and experiment with how different sounds are made.

You will need:
a rectangular box (a cereal box or shoebox)
rubber bands of different widths
a thumbtack
a wooden ruler, or a small block of wood
a quarter

1. Stretch the rubber bands around the box. Line them up from thinnest to thickest.
2. Press the thumbtack into the lower edge of the box (as shown).
3. Use the quarter to pluck the strings of your instrument. Listen to the sounds. Pay attention to the pitch of the different sounds.
4. Place the wooden ruler (or block) under the rubber bands (as shown) to make a bridge. Listen to the sounds.
5. Wind one of the rubber bands around the thumbtack. This will tighten the tension on that band. Listen to the sounds made as you tighten the band more and more.

Find out:

which rubber bands produce the highest-pitched sounds

how the sounds change as the bands get shorter

how the sounds are different

how the sounds differ as the rubber band gets tighter

which bands produce low pitches

...Tra la la ... .with a banjo on my knee.

What's making that horrible noise?

bridge
(ruler or wooden block)

strings
(rubber bands of various widths)

thumbtack
(to wind rubber bands around)

(cereal box, or shoe box)

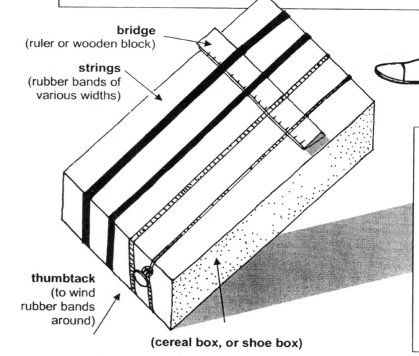

How it works:
As the rubber bands get thinner, shorter, or tighter, the pitch of the sound will be higher. This is because each of these makes the surface that is vibrating smaller, and results in faster vibrations. The faster the vibration, the higher the pitch. The wider, looser, longer rubber bands vibrate more slowly.

GET SHARP →

on Science Terms

Get Sharp on Science Terms

A

abrading – scouring of rock surfaces by a glacier

abrasion – process in which wind carries fine particles that scour rock surfaces

absolute dating – a method that gives the actual time of an event in Earth's history

absolute magnitude – the measure of a star's actual brightness

absorb – to take in or soak up

abyss – ocean depths of 2000-6000 meters

accelerate – to increase speed in movement

acceleration – the rate at which the velocity of an object changes

acid – a chemical substance that reacts with metals to release hydrogen; produces hydronium ions when dissolved in water

acid rain – pollution caused by waste compounds in the air forming weak acids in rain water

action-reaction pairs – a pair of forces which act on an object equally from opposite directions

active transport – movement of material by cells from areas of lower concentration to areas of higher concentration

active volcano – a volcano that erupts frequently

adaptation – a change in structure, function, or form that helps an organism adjust to its environment

adrenalin – hormone released by adrenal glands

aerobic exercise – exercise that strengthens the heart and lungs

aftershocks – tremors that follow an earthquake

air – the mixture of gases that makes up Earth's atmosphere

air mass – a body of air that has the same properties as the region over which it develops

air pressure – the pressure that air in the atmosphere exerts on everything on Earth's surface

algae – simple plants, without proper leaves, stems, or roots

alkali – a chemical substance that neutralizes acids

alternating current – current that changes direction

altitude – the height that something is above sea level

alveoli – tiny, thin-walled sacs in the lungs that are filled with air and surrounded by capillaries

ampere (amp) – the unit used to measure the flow of electrical current

amphibians – the class of vertebrates, including frogs, toads, and salamanders, that begins life in the water as tadpoles with gills and later develops lungs

amplitude – the greatest distance that the particles in a wave rise or fall from their rest position

anemometer – an instrument used to measure wind speed

angiosperm – a seed plant that produces seeds inside a fruit; a flowering plant

angle of incidence – angle at which incident waves strike an object

angle of reflection – angle at which reflected waves bounce off an object

antennae – a pair of movable, jointed sense organs on the heads of insects and other related organisms; used for taste, touch and smell

antibodies – proteins made by the body that destroy poisons created by germs

antibiotic – a substance produced by a living organism that slows down or stops the growth of bacteria

antiseptic – a substance that sterilizes and kills germs

aorta – largest artery in the body; carries blood away from the heart

aphelion – the point in an orbit where the body is farthest from the Sun

apparent magnitude – the brightness of a star as seen from Earth

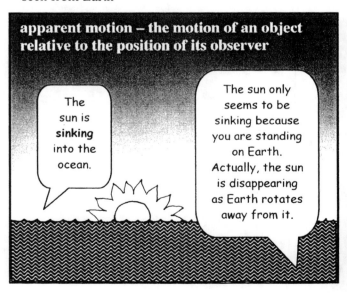

apparent motion – the motion of an object relative to the position of its observer

The sun is **sinking** into the ocean.

The sun only seems to be sinking because you are standing on Earth. Actually, the sun is disappearing as Earth rotates away from it.

aquifers – permeable rocks containing water

arachnids – the class of arthropods, including spiders and scorpions, which have four pairs of legs, no antennae, and which breathe through lung-like sacs or breathing tubes

arthropoda – the phylum of invertebrate animals with jointed legs and a segmented body (such as insects, crustaceans, and arachnids)

arteries – vessels carrying blood away from the heart

asteroids – numerous small planets orbiting between the orbits of Mars and Jupiter

asthenosphere – the part of Earth's mantle that is most fluid

astronomy – the study of the stars, planets and other heavenly bodies

atrium – one of two upper chambers of the heart

atmosphere – the gaseous mass that surrounds any star or planet

atom – a tiny particle of matter consisting of a nucleus that contains protons and neutrons and an electron cloud that contains electrons

atomic mass – the number of protons plus the number of neutrons in the nucleus of an element

atomic number – the number of protons in the nucleus of an atom; identifies the kind of atom

auditory canal – tube leading from outer ear to inner ear

auditory nerve – nerve that carries sound vibrations and messages to the brain

aurora australias – streams of light in the upper atmosphere near the South Pole

aurora borealis – streams of light in the upper atmosphere near the North Pole

axis – imaginary line around which an object spins

B

balanced forces – two forces acting on an object with equal strength from opposite directions

barometer – instrument that measures air pressure

base – a substance that increases the hydroxide ion concentration when added to water

bedrock – the solid rock found under soil

biochemistry – the branch of chemistry that deals with plants and animals and their life processes

biome – an extensive community of plants and animals whose makeup is determined by soil, climate, and other geographical features

biosphere – regions of Earth's land, water, and atmosphere inhabited by living things

bladder – organ that stores urine

boiling point – the temperature at which a liquid turns into a gas

bone marrow – soft tissue at the center of bone; place where blood cells are made

botany – the study of plants

bronchi – tubes in the respiratory system that branch off the trachea

bronchioles – smaller tubes into the lungs which branch off the bronchi

budding – a form of asexual reproduction in which a new organism grows from a bulge of tissue on the parent

buoyant force – the force with which a fluid pushes on objects

I believe that something other than a buoyant force is pushing up against me.

C

cambium – plant tissue that makes stems grow thicker

camouflage – body coloring that protects an organism

capillary – one of many tiny (microscopic) blood vessels which join small arteries to veins

carbohydrate – any of certain nutrients made of sugar or starch

carbon dioxide – a colorless, odorless gas that is used by green plants and some protists in photosynthesis and which is given off by all living things in respiration

carbohydrates – energy-rich compound that comes from foods

cartilage – rubbery protein that cushions movable joints

cardiac muscle – muscle in the heart

catalyst – a substance that speeds up chemical reactions, but is not changed by the reaction

cause – anything that brings about a result

Celsius scale – a temperature scale used in the metric system at which water freezes at 0 degrees

central nervous system – brain and spinal cord

centrifugal force – a force acting on an object spinning around another to push the object outward

centripetal force – a force on an object operating toward the center of a circular path

cerebellum – part of the brain that controls balance and voluntary muscle action

cerebrum – largest part of the brain; controls thinking and awareness

change – the process of becoming different

chemical change – a change in which atoms and molecules form or break chemical bonds

I'm pretty sure I've produced a chemical change here.

chemical energy – energy released during a chemical reaction

chemical equation – a description of a chemical reaction using symbols and formulas

chemical formula – the combination of chemical symbols used as a shorthand for the name of a compound

chemical property – a property that describes the behavior of a substance when it reacts with other substances

chemical reaction – a change that produces one or more new substances

chemical symbol – the shorthand way of writing the name of an element

chemistry – the study of matter, its composition and its changes

chlorophyll – the chemical in chloroplasts of plant cells that is needed for photosynthesis

chordata – the phylum of animals with an internal skeleton

chromosphere – the bright red layer of gas surrounding the Sun's photosphere

chromosomes – microscopic, rod-shaped bodies, which carry the genes that convey hereditary characteristics

chrysalis – pupa stage in insect development from the caterpillar stage

circuit – an unbroken path formed by electrical conductors

circulation – the movement of blood around the body due to pumping of the heart

classify – to put objects or processes in a group or category based on a common characteristic or group of characteristics

cleavage – breakage of a mineral along smooth, flat planes

climate – the average long-range weather of an area

cloud – tiny droplets of water grouped together in the atmosphere

cochlea – spiral-shaped ear structure; sound waves stimulate it to produce nerve impulses

cocoon – case of threads that an insect larva spins around itself

coefficient – a number that tells how many molecules of a substance are needed or produced in a reaction

coelenterata – the phylum of invertebrate animals with a central cavity and tentacles

colon – the last section of the large intestine

colloid – a heterogeneous mixture with particles of a size between that of a suspension and a solution

coma – the thick cloud of water and gases surrounding the nucleus of a comet

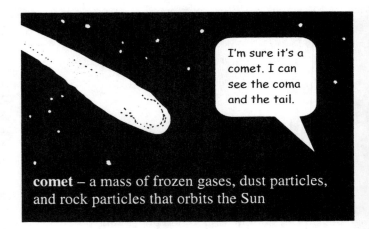

I'm sure it's a comet. I can see the coma and the tail.

comet – a mass of frozen gases, dust particles, and rock particles that orbits the Sun

commensalism – a relationship in which one organism lives on another without harming it

communicate – tell or show others the steps and results of an experiment

community – all the plants and animals that live together in a habitat

compound – a substance made up of two or more elements

222

compound machine – a machine made by combining two or more simple machines

compressional wave – a wave in which matter vibrates in the same direction that the wave moves

conductor – a material that transmits or carries electricity or heat

conduction – the transfer of heat from particle to particle when two substances come in contact

conglomerate – sedimentary rock made of pebbles and gravel cemented together by clay

conjugation – a method of reproduction that results in a zygospore

conifer – a seed plant that produces seeds in cones

conservation – all efforts to protect, replace, or make careful use of resources

conservation of energy – the principle that energy cannot be made or destroyed, but only changed in form, and that the total energy in a physical system cannot be increased or diminished

constancy – a state characterized by a lack of variation

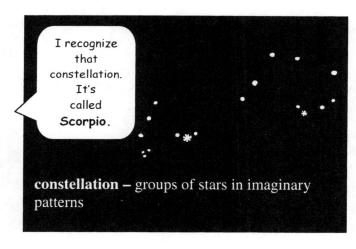

I recognize that constellation. It's called **Scorpio**.

constellation – groups of stars in imaginary patterns

convection – the transfer of heat by movement of matter

convection current – the movement of material within a fluid caused by uneven temperature; the upward movement of warm air and the downward movement of cool air

convergent boundaries – boundaries between plates in Earth's surface where plates collide

core – the center of the Sun; the outer atmosphere of the Sun

cornea – tough, protective outer covering of the eye

coulomb – the charge on 6.24 billion billion electrons

crater – basin-like depression at the top of a collapsed volcano

creep – the slow, downslope movement of a mass of Earth material

crust – the outermost layer of Earth, extending to a depth of about 35 km

crustaceans – the class of arthropods, including lobster, crabs, and shrimps, that usually live in the water, breathe through gills, and have a hard outer shell and jointed appendages

crystal – a solidified form of a substance in which the atoms or molecules are arranged in a definite pattern

cycle – a series of events or operations that occurs regularly and usually leads back to the starting point

cytology – the study of cells

D

deceleration – negative acceleration; the rate of change in velocity when the velocity decreases

decibel – the unit for measuring the loudness of sound

decomposer – an organism that helps another organism to decay

decomposition – weathering by chemical processes

deep water wave – a wave moving in water that is deeper than one-half its wavelength

deflation – process in which wind picks up loose material from the ground surface and moves it

delta – fan-shaped deposit of sediment at the mouth of a river

density – the ratio of the mass of an object to its volume

density currents – currents formed by the movement of more dense water toward an area of less dense water

dermis – second layer of the skin

diaphragm – muscle between chest cavity and abdomen; assists with breathing

dicotyledon – a flowering plant that has two leaves (cotyledons) inside the seed; leaves have a network of veins

diffusion – the movement of particles in solution from an area of greater concentration to an area of lower concentration

direct current – electric current that moves in one direction only

disintegration – weathering by physical processes

diurnal – an event occurring once a day; usually referring to tides

divergent boundaries – boundaries between plates in Earth's surface where two plates pull apart

---

223

dominant species – a species that gets more of the resources in an environment and therefore survives better

DNA – (deoxyribonucleic acid) the acid in chromosomes that carries genetic information

dormant volcano – volcano that is inactive but could become active again

Doppler effect – a change in the frequency of sound waves due to the movement of the listener or the object making the sound

dunes – piles of sand dropped when sand-carrying wind meets an obstacle

E

earthquake – vibrations in Earth's crust caused by the movement or breaking of rock in the surface

echinodermata – the phylum of marine invertebrate animals with a water vascular system and usually a hard, spiny skeleton and radial body (starfish, sea urchins, etc.)

eclipse – the passing of one object into the shadow of another object

ecology – the study of the relationship between plants, animals, and their environment

ecosystem – a system consisting of a community of animals, plants, and bacteria and their interrelated physical and chemical environment

effect – a result; an event or situation that follows from a cause

effort force – the amount of force that must be applied to a machine

electric charge – a positive or negative charge resulting when an object has too many or too few electrons

electric current – a steady flow of electrons through a conductor

electric energy – the energy used by an electrical device

electric power – the rate at which a device changes electricity into another form of energy

electricity – energy resulting from the flow of electrons

electromagnet – a coil of wire with current flowing through it that becomes a magnet

electromagnetic waves – transverse waves that travel at the speed of light in a vacuum

electromagnetic spectrum – an arrangement of the different electromagnetic waves according to their lengths

electron – a negatively charged atomic particle

element – a substance made up of only one kind of atom

elevation – the distance of a point above or below sea level

embryo – a young plant or fertilized cell

emulsion – a suspension of two liquids

endocrinology – the study of the endocrine system

energy-matter – the close relationships and interactions between matter and energy

evaporation – the change in state from solid or liquid to a gas

environment – the part of the biosphere surrounding a particular organism

epidermis – protective outer layer of skin

epiglottis – protective flap at the top of the trachea

equinox – a day when the hours of darkness and daylight are the same length

erosion – the moving of weathered particles on Earth's surface

Eustachian tubes – tubes that equalize pressure in the ear

exoskeleton – the hard outer covering protecting inner organs of an arthropod

experiment – a planned series of steps designed to find an answer to a question

external fertilization – the joining of a sperm and egg outside the body of the animal

extinct – volcano that is inactive for a very long time; form of life no longer living

F

Fahrenheit – the temperature scale in which the freezing point of water is 32 degrees and the boiling point of water is 212 degrees

fault – a fracture in a rock along which movement has taken place

fertilization – the joining of nuclei of the male and female reproductive cells

fission – splitting or breaking into part

flagellum – the whip-like tail on some simple-celled animals that helps in movement

floodplain – an area formed by sediment deposited outside the river bed during flooding

fluid – a substance that flows, such as a liquid or gas

food chain – the path of food energy from one organism to another in an ecosystem

food web – a complete network of food chains

force – a push or a pull acting on something

form – the shape or other physical characteristics of an object, organism, or system

form & function – the relationship between the shape (form) of an object, organism, or system and its operation (function); usually a relationship where the function is dependent upon the form

fracture – break in rock or bones; irregular breakage of a mineral

fragmentation – a form of asexual reproduction where an animal divides into two or more pieces

frequency – the number of waves that pass a given point per second

freezing point – the temperature at which a substance changes from a liquid or gas to a solid

friction – a force that opposes motion between two surfaces that touch each other

front – a boundary where air masses meet

fulcrum – the point on which a lever is supported

function – the operation of an object, organism, or system

full Moon – Moon phase occurring when Earth is between the Sun and the Moon, with the Sun shining on the Moon so that it is visible from Earth

fungi – a kingdom of plant-like organisms that are parasites on living organisms or feed upon dead organic material and which lack true roots, chlorophyll, stems, and leaves, and reproduce by means of spores

fusion – the joining together of parts into a whole

G

galaxy – a large grouping of millions of stars, planets, dust, and gases in outer space

gallbladder – organ that produces bile

galvanometer – a tool used for measuring very small electrical currents

gas – the form of matter that has no definite shape or volume

gemstones – a mineral or petrified substance that can be used as a gem when cut and polished

genes – units of inheritance passed from parents to offspring

genetics – the study of heredity

Genetics explains why you tend to look like you whereas I tend to look like me.

germination – a process where seeds begin to grow from an embryo into a seedling

geologic time scale – a history of the Earth based on observations of rocks and fossils

geyser – a spring from which boiling water and steam shoot into the air at intervals

glacier – a moving river of ice and snow

glacial flow – the movement of glacial ice

gravity – the force of attraction that exists between all objects in the universe

groundwater – underground water

gymnosperms – a large class of seed plants which have the ovules borne on open scales (usually in cones) and which lack true vessels in the woody tissue (pines, spruces, cedars, etc.)

H

habitat – the type of environment suitable for an organism; native environment

hardness – a mineral's resistance to being scratched

hematology – the study of blood

heredity – the passing on of traits from parents to offspring by means of genes in the chromosomes

heterogeneous mixture – a mixture in which the composition is not the same throughout

hibernation – an animal behavior that involves a long period of rest or inactivity, usually in winter

homeostasis – the tendency of an organism toward balance

homogeneous mixture – a mixture in which particles of one substance are spread evenly throughout another substance

humidity – the ability of air to hold water

hydrosphere – all of the water on the face of the Earth

hypothesize – to make an assumption or a careful guess in order to test it further

I

incident waves – waves that strike an object

indicator – a device or substance that can be used to determine the acidity or alkalinity of a solution

inertia – the property of matter to resist changes in motion

infer – to draw a conclusion based on facts or information gained from an inquiry

igneous rocks – rocks formed from the cooling of hot, molten magma

inherited traits – traits that are passed on from parents to offspring

insoluble – that which cannot be dissolved

It's too bad that my report card is insoluble in water.

insulin – hormone that controls the amount of sugar in the bloodstream and the storage of sugar in the liver

intensity – the loudness of a sound

internal fertilization – joining of a sperm and egg that takes place inside the body of the female animal

interpret – to explain or tell the meaning of the results of an investigation

ion – an electrically charged atom that has lost or gained one or more electrons in a chemical reaction

iris – colored part of the eye; muscle that expands and contracts to let light into the eye

insulator – a substance that does not conduct heat or electricity well

investigation – an experiment or other organized plan for answering a question

J

jet stream – very rapid winds that move around the Earth from west to east at a high altitude

joints – place where a bone joins to another

joule – the unit used for measuring energy or work

K

kidneys – organs that remove waste from the blood

kinetic energy – energy of motion

kingdom – the major classification category of living organisms

L

landslides – quick movement of large amounts of material downhill

larva – the free-living, immature form of any animal that changes structurally when it becomes an adult; the second stage of insect development

larynx – voice box; part of the air passage between the mouth and nose and the trachea

lens – curved, transparent object that changes the direction of light; transparent disc in the eye that bends light to focus images

ligaments – strong, flexible fibers that hold bones together and stretch to allow bending at joints

liquid – the form of matter that has a definite volume but no definite shape

lithosphere – the outer rocky region of Earth that includes the crust and the rigid upper layer of the mantle

liver – organ that cleans wastes from the blood and stores useful substances

loess – a windblown deposit of fine dust particles gathered from deserts, dry riverbeds, or old glacial lakebeds

luminosity – the rate at which a star pours out energy

lunar eclipse – the partial or total obscuring of the Moon when Earth comes directly between the Sun and the Moon

M

machine – a device that changes the amount or direction of force that must be exerted to do work

magma – liquid or molten rock deep inside the Earth

magnet – a substance or objects that attracts other magnetic objects to itself

magnetic field – a region around a magnet where magnetic force acts

magnetic poles – opposite ends of a magnet; place where magnetic force is strongest

main sequence period – the main life period of a star

mammal – a warm-blooded vertebrate that produces milk to feed its young; most develop inside the female's body before birth

mantle – the thick layer of the Earth between the crust and the core

marine – something that lives in the sea or is formed by the sea

mass – the amount of matter in an object

matter – anything that has mass and takes up space

measure – to compare an object or amount to a standard quantity in order to find out an amount or an extent

mechanical advantage – the ratio of the force that performs the useful work of a machine to the force that is applied to the machine

mechanical energy – potential and kinetic energy combined in lifting, stretching, or bending

medulla – the part of the brain at the base of the skull; controls involuntary muscle activities

meridian – imaginary lines running from Earth's north pole to south pole

metamorphic rocks – rocks that have been changed by heat and pressure

metabolism – the sum total of all the chemical processes and changes in an organism

metamorphosis – a process of going through structural changes to take the shape of an adult animal

meteor – the flash of light that occurs when a meteoroid is heated by its entry into the Earth's atmosphere; a shooting or falling star

meteoroid – any of the small, solid bodies that travel through outer space and are seen as meteors when they enter Earth's atmosphere

meteorite – the part of a meteoroid that falls to the Earth's surface

migration – an animal behavior that involves moving long distances to reproduce, mate, raise young, or find food

mineral – a naturally occurring, inorganic, crystalline solid with a definite chemical make-up

mitosis – the division of a cell nucleus

mixture – a substance containing two or more substances which are not in fixed proportions; the substances do not lose their individual characteristics when combined

molecule – the smallest particle of an element or compound that can exist in the free state and still retain the characteristics of the element or compound

model – a structure that visually represents real objects or events

mollusca – the phylum of invertebrates characterized by a soft, unsegmented body (often closed in a shell), and which usually has gills and a foot (oysters, snails, clams, etc.)

molting – a process by which an animal sheds its outer covering

Moment Magnitude Scale – a means for measuring the magnitude of earthquakes

momentum – the force produced by a moving body; the product of an object's mass and its velocity

monocotyledon – a flowering plant that has only one cotyledon (seed leaf) in its seed; has parallel veins

moraine – a huge mass of rocks, gravel, sand, and clay that has been deposited by a glacier

motion – a change in position of matter

mudflows – rapid movement of soil from weathering mixed with rain down a slope

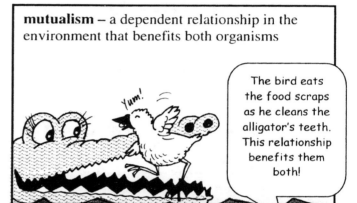

mutualism – a dependent relationship in the environment that benefits both organisms

The bird eats the food scraps as he cleans the alligator's teeth. This relationship benefits them both!

N

neap tides – low tides that occur when the Sun, Earth, and Moon form a right angle

nebulae – clouds of dust and gas where stars are born

negative charge – the charge of an atom having an excess of electrons

net force – a force resulting when the forces acting on an object are unbalanced

neuron – a nerve cell

neutral – neither positively nor negatively charged; neither acidic nor basic

neutron – a neutral atomic particle

new Moon – phase of the Moon in which the side of the Moon facing Earth is dark

non-electrolyte – a substance that will not make water conduct electricity

nonvascular plants – plants without vessels

nuclear energy – energy stored in the nucleus of atoms and released when atoms are split or fused

nucleus – the center of an atom which contains protons and neutrons; the solid part of a comet

nutrient – a chemical substance found in foods which is necessary for the growth or development of an organism

nutrition – the process of eating, digesting, and absorbing food; the study of healthy diets

nymph – the young stage in the development of insects that experience incomplete metamorphosis

O

observe – to recognize and note facts or occurrences; to watch carefully

ohm – the unit used for measuring resistance in an electrical conductor

orbit – the path of one object in free-fall around another object in space

order – the predictable behavior of objects, units of matter, events, organisms, or systems

offspring – a new organism produced by a living thing

optic nerve – nerve that carries messages from the eyes to the brain

orbit – the path a body follows in revolving around another body

organ – a group of specialized tissues that work together to perform a special function

organism – a living thing

organization – the arrangement of independent items, objects, organisms, units of matter or systems, jointed into a whole system or structure

orogeny – the processes that build mountains

osmosis – diffusion of water through a membrane

ovaries – female organs that produce eggs

ovum – a female reproductive cell; an egg

oxidation – the union of a substance with oxygen; the process of increasing the positive capacity of an element or the negative capacity of an element to combine with another to form molecules; the process of removing electrons from atoms or ions

P

pancreas – organ that produces insulin which controls sugar in the blood

parallel circuit – a circuit with two or more separate paths for current to follow

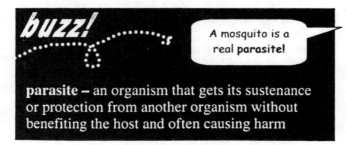

A mosquito is a real parasite!

parasite – an organism that gets its sustenance or protection from another organism without benefiting the host and often causing harm

parathyroids – glands that regulate the balance of calcium in the bones and blood

partial eclipse – an eclipse in which the body (Moon or Sun) is only partially covered

penumbra – a partial shadow formed during an eclipse

perihelion – the point in an orbit where the body is closest to the Sun

period – a subdivision of a geologic era; the time taken for a cycle of events to take place; rows of elements in the periodic table

periodic table – an arrangement of elements in order of their atomic numbers

peripheral nervous system – nerves not including those in the brain and spinal cord

petal – the colored outer part of a flower

pH scale – a range of numbers used to tell the acidity or alkalinity of a solution

phases – any of the recurring stages of changes in the appearance of the Moon or a planet

phloem – plant tissues with tube-like cells that transport food from leaves to other plant parts

photon – a particle of radiation

photosynthesis – the process in which green plants use the Sun's energy to produce food

phylum – the largest classification category in a kingdom

Oops! This is going to be a costly physical change!

physical change – a change in which chemical bonds are not formed or broken and no new substance is produced

physical property – a property that distinguishes one type of matter from another and can be observed without changing the identity of the substance

physics – the study of different forms of energy

pitch – a quality of sound (highness or lowness) determined by wave frequency

pituitary – master gland of the body; produces growth hormones

plain – a large, flat area with an elevation that differs little from that of the surrounding area

planet – an object in space that reflects light from a nearby star around which it revolves

228

plasma – fluid in which blood cells travel

plasmolysis – shrinking of cytoplasm in cells due to water loss

platelets – substances in the blood that produce clots

plates – huge sections of Earth's crust which lie beneath oceans and continents over a layer of molten rock in the mantle

plate tectonics – a theory that explains movements of continents and changes in Earth's crust caused by internal forces

plucking – the process of picking up fragments by a glacier and carrying them along

podiatry – the medical treatment of feet

pollen – the yellow, powder-like male reproductive cells formed in the anther of the stamen of a flower

Could that insect be allergic to pollen?

pollination – the movement of pollen from a stamen to the pistil of the flower; insects often assist with pollination

population – the number of individuals of one species in a community

porifera – phylum of chambered animals (sponges) that live in water

positive charge – the charge of an atom having an excess of protons

potential difference – the difference in potential energy between the electrons in one place and the electrons in another

potential energy – energy due to position or condition

power – the rate of doing work

precipitate – an undissolved solid that usually sinks to the bottom of a mixture

precipitation – the falling of water or ice formed by condensation

predator – any animal that hunts and eats other animals

predict – to foretell what is likely to happen based on an observation or experiment

prehistoric – before recorded history

pressure – a pushing or squeezing force

prey – animals that are hunted and eaten as food

prominences – fingers of flame that surge from the Sun's chromosphere

property – a quality that describes or characterizes an object

proteins – organic compounds made of amino acids; important body nutrients necessary for life and growth in all organisms

protists – kingdom of simple organisms, mostly one-celled; most do not make their own food

proton – a positively charged particle found in the nucleus of an atom

protoplasm – the essential living material of all animal and plant cells

protozoa – the phylum of mostly microscopic animals made up of a single cell or group of identical cells and living mainly in water; many are parasites

pulmonary arteries – vessels that carry oxygen-rich blood from the lungs

pulmonary veins – vessels that carry carbon dioxide-laden blood to the lungs

pupil – tiny opening in the eye that lets light in

pulsar – rapidly rotating neutron star that gives out a beam of radiation which looks like a pulse

Q

quarantine – to separate an organism from others

quark – an subatomic particle

quasar – a quasi-stellar radio source; a starlike object that emits energy in waves

R

radiation – the transfer of energy by electromagnetic waves

receptors – nerve cells that receive impulses

reflection – bouncing off of waves or rays from an object

refraction – the bending of light as it passes from one medium to another

relative dating – a method used to put events of Earth's history in sequence in relationship to each other

reproduction – the process by which organisms produce offspring

reptiles – cold-blooded animals with scales; live mostly on land and breathe air

resistance – any opposition that slows something down or prevents movement

resistance force – the force exerted by a machine

respiration – the process in which cells release energy from food

response – a change in an organism's behavior as a result of a stimulus

retina – screen of light-sensitive receptor cells which receive images in the back of the eye

revolve – to travel in an orbit around an object

Earthquake!

I guess I should have bought a Richter scale instead.

Richter scale – a means for measuring the magnitude of earthquakes

riverbed – the path through which a river flows

river load – the material carried by a river

revolution – the movement of a body (or object) around another body (or object)

rotation – the turning or spinning of an object on an axis

runoff – water from precipitation that flows across Earth's surface and eventually returns to lakes, rivers, and oceans

S

salt – the substance that results when the right quantities of an acid and a base neutralize each other

satellite – a small planet that revolves around a larger one; a man-made object put into orbit around some heavenly body

saturated solution – a solution that holds all the solute it can at a given temperature

scavengers – animals that feed on dead organisms

scientific inquiry – a way of doing investigations and looking for explanations about happenings in the physical world; a series of steps generally followed in looking for answers to questions in science

seed – a ripe, fertilized ovule that will develop into a plant under suitable conditions

seedling – a young plant

sedimentary rocks – rocks formed by the cementing together of materials

semicircular canals – canals in the ear containing fluid; help keep balance

sense – a power that allows animals to be aware of their surroundings

seismic waves – vibrations set up by earthquakes

seismograph – an instrument that measures movements in the Earth's crust

sepals – the green, leaf-like structures that surround the bottom of flowers

series circuit – a circuit with only one path for current to follow

shallow water wave – a wave in water shallower than one-half its wavelength

shooting star – a briefly visible meteor

shoreline – the boundary where land meets the water of the ocean or lake

shore zone – area along the shore between the point of high tide and low tide

simple machine – a machine consisting of only one part

slump – a curved scar left when layers of rock slip downslope

smooth muscle – involuntary muscle present in walls of many internal organs

soil – a mixture of decayed organic material, weathered rock, air, and water, that covers much of the land on Earth

soil horizons – layers or regions of soil

solar eclipse – an eclipse that occurs when Earth is in the Moon's shadow

solar energy – energy that is trapped from the Sun

solar flares – sudden increases in brightness of the Sun's chromosphere

solid – a form of matter which has definite size and shape

solstice – a day when one of Earth's poles is tilted directly toward or away from the Sun

solubility – the amount of a substance that will dissolve in a specific amount of another substance at a given temperature

solute – the substance being dissolved in a solution

solution – a homogeneous mixture with tiny particles of a substance spread throughout another substance; particles cannot be filtered or settled out

solvent – the substance in which a solute is dissolved in solution

species – the smallest category in the kingdom in which only one kind of organism is classified

speed – the rate of motion of a body or object

sperm – male reproductive cell

spinal cord – thick cord of nerves that runs from the brain through the vertebrae

spring tide – a tide that occurs when the Sun, Moon, and Earth are aligned

spore – reproductive cell; grows into an organism

stamen – the male reproductive organ of an angiosperm

static electricity – electricity produced by charged bodies; charge built up in one place

stimulus – something happening in the environment that affects the behavior of an organism

stirrup, hammer, anvil – three bones of the inner ear that carry vibrations from sound to the auditory nerve

stratosphere – the second layer of the atmosphere (above the troposphere) which extends six to fifteen miles above the Earth's surface and where the temperature is fairly constant

subatomic particles – particles which make up smaller than atoms (protons, neutrons, electrons, quarks)

substance – a chemical element or compound with a known composition

sunspots – dark spots on the Sun where the temperature is cooler

supersaturated solution – a solution that holds more solute than normal at a given temperature

suspension – a cloudy mixture of two or more substances that settles on standing

striated muscle – voluntary muscle made of bands called striations (also called skeletal muscle)

switch – a device on a circuit that can open and close the pathway for flowing current

symbol – the shorthand way to write the name of an element

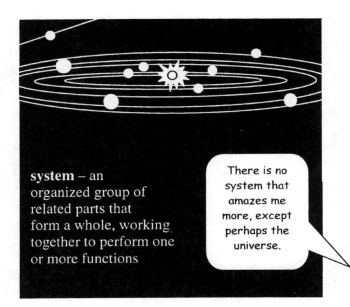

system – an organized group of related parts that form a whole, working together to perform one or more functions

There is no system that amazes me more, except perhaps the universe.

T

talus – eroded material that collects at the foot of a steep slope or cliff

taste buds – groups of receptor cells on the tongue sensitive to substances dissolved in saliva

telescope – an instrument which makes distant objects appear closer and larger

temperature – the measure of average kinetic energy of the particles in a material

tendons – strong bands of tissue that attach the ends of muscle fibers to bones

territoriality – actions designed to protect or defend a certain geographical area

testes – sperm-producing organ of the male reproductive system

thermal energy – the total energy in all the particles of an object; the amount depends on its temperature

thyroid – gland that controls the rate at which food is used in the body

tides – shallow water waves caused by the gravitational attraction among Earth, Moon, and Sun

till – unsorted, unlayered glacial deposit of boulders, sand, and clay

tissue – groups of similar cells that specialize to do a particular job in the body

tone quality – the difference among sounds that have the same pitch and loudness

topsoil – the highest, richest layer of soil (Horizon A)

total eclipse – an eclipse in which the Moon or Sun is entirely obscured from view for a period of time

trachea – windpipe; tube carrying air from the mouth to the lungs

trait – a characteristic

transform faults – boundaries where plates slide or scrape each other

translucent – material that transmits light but does not allow you to see through it clearly

transparent – having the property of transmitting or passing light

transpiration – the process through which organisms lose water

transverse wave – a wave in which matter moves at right angles to the direction the wave is traveling

tributaries – streams and small rivers that join into a large river

tropism – the response of a plant to a stimulus

troposphere – layer of atmosphere nearest Earth; contains about 90% of the gases in atmosphere

U

ultrasound – sound waves of a high frequency (above 20,000 Hz)

umbra – an inner, complete shadow formed during an eclipse

universe – the total of everything that exists in space

urethra – tube that carries urine out of the bladder and out of the body

ureters – tubes that carry urine from the kidneys to the bladder

uterus – female reproductive organ which holds a growing fetus

V

vacuum – the absence of matter

vapor – the gaseous form of a substance that is usually a solid or liquid at ordinary temperatures

vascular plants – plants with vessels

veins – vessels that carry blood toward the heart

velocity – the speed and direction of a moving object; the distance traveled by any point on a wave in one second

vent – tube-like passage through which magma and gases escape from a volcano

burp!

I guess every volcano has to vent now and then.

ventricle – one of two lower heart chambers

vertebra – one of the small bones that makes up the spinal column or backbone

vertebrates – animals with backbones

vibrations – rapid back and forth movements; the up and down movement of a wave

viscosity – the property of a liquid that describes how it pours

volt – the unit used for measuring the push of electricity through a conductor

W

waste – unwanted products of industrial processes, digestion, or respiration

water table – level in the ground below which water collects and cannot drain down any farther

watt – the unit for measuring electrical power

waves – ocean movements in which water rises and falls

weathering – the process by which surface rocks and other materials are broken down by wind, water, and ice

wedge – a simple machine that is an inclined plane with either one or two sloping slides

weight – the force of gravity that a planet exerts on objects resting on the surface

wheel and axle – a simple machine that is a variation of a lever; consists of a large wheel fixed to a smaller wheel that rotate together

wind – movements of air parallel to the Earth's surface

work – the transfer of energy as a result of motion

wormhole – theoretical tunnels that link one part of space-time with another

X

X-ray – an invisible form of radiation with a very short wavelength; can pass through many materials that stop other rays, such as light rays

xylem – the tissue in plant roots and stems that supports the plant and carries water and nutrients to stems and leaves

Y

yeast – a type of fungus that feeds on sugary substances and reproduces by budding

Z

zone – one of the areas of Earth's surface that has a particular type of climate

zoology – the study of animals

zygote – a cell formed by fertilization

Index

Better Grades & Higher Test Scores / SCIENCE
Copyright ©2003 by Incentive Publications, Inc., Nashville, TN

O

oasis, 117
observation, 63, 64
ocean, 114, 125-128
    currents, 126
    floor, 125
    movements, 126-127
    shoreline features, 128
occluded front, 133
Oersted, Hans Christian, 211
ohm, 208
Ohm's Law, 55
Old Faithful, 122
olfactory lobe, 171
opague, 214
operational definitions, 65
optic nerve, 171
orbits,
    of comets, 85
    of planets, 75, 78-79
organs, 164-174
organic compounds, 188
organisms, 138
organization, 18-28
osmosis, 141
outer planets, 79, 82-83
outwash, 118
ovary, 150-151, 167, 174
ovule, 150-151
oxbow lake, 121
oxygen, 165
ozone layer, 162

P

pancreas, 166-167
Pangaea, 105
parallel circuit, 209
parasite, 161
parathyroid gland, 167
parthenogenesis, 154
partial eclipse, 87, 88
Pascal's Principle, 55, 185
Pasteur, Louis, 61
peak, 114-115
peninsula, 114-115
penis, 174
penumbra, 87, 88
perihelion, 75
perinneals, 148
periods, 113
periodic table, 51, 186-187
periosteum, 169
peripheral nervous system, 170
permeable rocks, 122
petiole, 148
pH, 195-196
pharynx, 165, 166
phloem, 148
photon, 213
photoperiodism, 147
phototropism, 147
photosynthesis, 146
physical changes, 197
physical properties, 184
physical sciences, 56
piedmont glaciers, 118

Pinckney, Elizabeth, 61
pistil, 151
pitch, 216, 217
pituitary gland, 167
plain, 114-115
planetoids, 84
planets, 74-75, 78-83
plantae, 143
plants, 138-141, 143, 152-157
    behaviors, 147
    classification, 144-145
    processes, 146
    nonvascular, 144
    structure, 139, 148
    tissues, 148
    vascular, 145
plasma, 172
plasmolysis, 141
Plate Tectonics Theory, 54, 104
plateau, 114-115
platelets, 172, 177
platyhelminths, 152
plucking, 118
Pluto, 79, 83, 98
polar easterlies, 130
Polaris, 94
poles, 210-211
pollen, 150-151
pollination, 150-151
pollution, 162-163
ponds, 124
populations, 160
poriferan, 152
potential difference, 207
potential energy, 202
power, 35, 205
precipitation, 132-135
predator, 161
predict, 37, 67
preparing for tests, 44-45
pressure,
    atmospheric, 129, 130, 199
    of fluids, 185
    of gases, 185
prey, 161
primary consumer, 160
primary waves, 106
principles, 55
processes,
    classifying, 65
    communicating, 67
    comparing, 64
    controlling variables, 65
    defining operationally, 65
    experimenting, 65
    formulating models, 66
    hypothesizing, 64
    inferring, 67
    interpreting data, 67
    life, 140-141
    observing, 64
    plant, 146
    predicting, 67
    questioning, 64
    recording data, 66
    scientific, 64-67

    summarizing data, 66
    using math, 66
producers, 160
prostate gland, 174
protein, 178
protista, 143
protons, 182
Proxima Centauri, 92
protostars, 94
Ptolemy, 51, 98
pulsars, 93, 95, 99
pupil, 171
Pythagoras, 61

Q

Quark Theory, 54
quasars, 91, 95, 99
questioning, 63, 64

R

radiant energy, 202
radiation, 204, 213-215
rain, 135
reading, 41
reason, 39
recall, 36
recessive genes, 175
recognize cause and effect, 37
recording data, 66
recycling, 163
rectum, 166
red dwarves, 94
red giants, 93-94
reflected waves, 214
reflection, 214
reforestation, 163
refraction, 214
relative dating, 113
relative humidity, 132
relativity, theory of, 54
renewable resources, 162-163
reproduction, 138, 141, 149-150, 154
    asexual, 149, 154
    budding, 154
    external fertilization, 154
    fragmentation, 154
    in animals, 154-155
    in plants, 149-151
    internal fertilization, 154
    metamorphosis, 155
    parthenogenesis, 154
    sexual, 150, 154
reptiles, 153
reserves, 163
resistance, 207, 208
resistance force, 205
resources, 162-163
respiration, 141, 146
respiratory system, 165
response, 138, 147
retina, 171
revolution, 136
rhizomes, 149
ribosomes, 139
Richter Scale, 106